Digital Signal Processing System Design: LabVIEW-Based Hybrid Programming

Nasser Kehtarnavaz

Digital Signal Processing System Design: LabVIEW-Based Hybrid Programming

by Nasser Kehtarnavaz
University of Texas at Dallas

With laboratory contributions by Namjin Kim
and Qingzhong Peng

AMSTERDAM • BOSTON • HEIDELBERG • LONDON • NEW YORK • OXFORD
PARIS • SAN DIEGO • SAN FRANCISCO • SINGAPORE • SYDNEY • TOKYO
Academic Press is an imprint of Elsevier

ACADEMIC
PRESS

Academic Press is an imprint of Elsevier
30 Corporate Drive, Suite 400, Burlington, MA 01803, USA
525 B Street, Suite 1900, San Diego, California 92101-4495, USA
84 Theobald's Road, London WC1X 8RR, UK

Library of Congress Cataloging-in-Publication Data
Application Submitted

British Library Cataloguing-in-Publication Data
A catalogue record for this book is available from the British Library.

ISBN: 978-0-12-374490-6

For information on all Academic Press publications
visit our Web site at www.books.elsevier.com

Printed in the United States of America
Transferred to Digital Printing in 2014

Contents

Contents

Preface

The previous edition of this book, titled *Digital Signal Processing System-Level Design Using LabVIEW*, showed how LabVIEW™ graphical programming can be used to build and analyze digital signal processing (DSP) systems in an interactive manner and in relatively shorter times as compared to text-based programming. The motivation for writing the previous edition was derived from the observation that many students taking DSP lab courses, in particular at the undergraduate level, often struggle and spend a fair amount of their time debugging C and MATLAB® codes in lab sessions instead of placing more effort into analyzing and thus understanding signal processing systems.

In this second edition of the book, graphical and textual programming are combined to provide a hybrid programming approach toward achieving a more effective mechanism to build and analyze DSP systems. Textual programming and graphical programming have their own merits and demerits from a programming point of view. In general, math operations are found to be easier to code in textual mode. For example, MATLAB provides a rich set of built-in functions for performing signal processing vector and matrix-based math operations. On the other hand, graphical programming offers an easy-to-build interactive and visualization environment and a more intuitive approach toward building signal processing systems.

In an effort to bring together the preferred features of textual and graphical programming, the labs in the previous edition have been redesigned by incorporating MATLAB code blocks or modules into the LabVIEW graphical programming environment via its new MathScripting feature. In other words, the coding for math-oriented modules is now done using M-files, while interactivity, visualization, and modularity are maintained by using LabVIEW.

In addition to the hybrid programming approach adopted in this second edition, the labs have been redesigned based on the latest release of LabVIEW (LabVIEW 8.5) at the time of this writing instead of LabVIEW 7.1, which was utilized in the first edition.

I would like to express my appreciation and gratitude to National Instruments, in particular the Academic Marketing Division, for their support of this book.

Nasser Kehtarnavaz
December 2007

What's On the Companion Website?

- The accompanying companion website includes all the lab files discussed throughout the book. These files are placed in corresponding folders as follows:

  - Lab01: Getting Familiar with LabVIEW: Part I

  - Lab02: Getting Familiar with LabVIEW: Part II

  - Lab03: Sampling, Quantization, and Reconstruction

  - Lab04: FIR/IIR Filtering System Design

  - Lab05: Data Type and Scaling

  - Lab06: Adaptive Filtering Systems

  - Lab07: FFT, STFT, and DWT

  - Lab08: Getting Familiar with Code Composer Studio

  - Lab09: DSP Integration Examples

  - Lab10: Hybrid Programming of Dual Tone Multi-Frequency System

  - Lab11: Hybrid Programming of 4-QAM Modem System

  - Lab12: Hybrid Programming of Cochlear Implant Simulator System

- To run the lab files, the National Instruments LabVIEW 8.5 is used and assumed installed. The lab files need to be copied into the folder "C:\LabVIEW Labs\", as shown in the following figure.

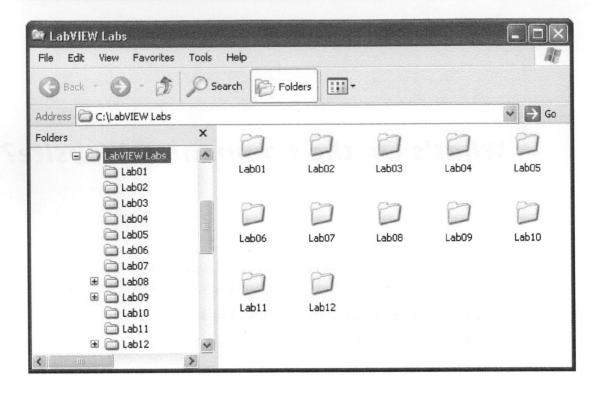

- For Lab 8 and Lab 9, the Texas Instruments Code Composer Studio™ (CCStudio) version 3.0 is used and assumed installed in the folder "C:\CCStudio\". The subfolders correspond to the following DSP platforms:

  - DSK 6416

  - DSK 6713

  - Simulator (configured as DSK6713 as shown in the following figure)

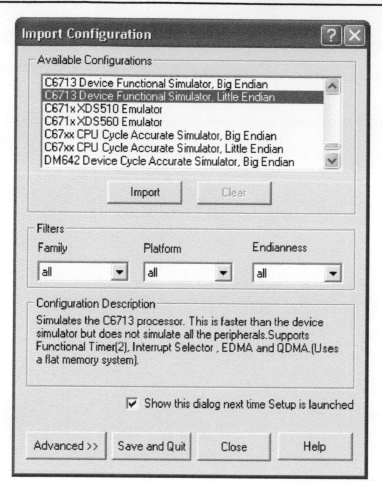

CHAPTER 1

Introduction

The field of digital signal processing (DSP) has experienced a considerable growth in the past two decades, primarily due to the availability and advancements in digital signal processors (also called DSPs). Nowadays, DSP systems such as cell phones and high-speed modems have become an integral part of our lives.

In general, sensors generate analog signals in response to various physical phenomena that occur in an analog manner (i.e., in continuous-time and amplitude). Processing of signals can be done either in analog or digital domain. To perform the processing of an analog signal in digital domain, it is required that a digital signal is formed by sampling and quantizing (digitizing) the analog signal. Hence, in contrast to an analog signal, a digital signal is discrete in both time and amplitude. The digitization process is achieved via an analog-to-digital (A/D) converter. The field of DSP involves the manipulation of digital signals in order to modify their characteristics or to extract useful information from them.

There are many reasons why one wishes to process an analog signal in a digital fashion by converting it into a digital signal. The main reason is that digital processing offers programmability, which means the same processor hardware can be used for many different applications by simply changing the code residing in memory. Another reason is that digital circuits provide a more stable and tolerant output than analog circuits—for instance, when subjected to temperature changes. In addition, the advantage of operating in digital domain may be intrinsic. For example, a linear phase filter or a steep-cutoff notch filter can be easily realized by using digital signal processing techniques, and many adaptive systems are achievable in a practical product only via digital manipulation of signals. In essence, digital representation (0's and 1's) allows voice, audio, image, and video data to be treated the same for error-tolerant digital transmission and storage purposes.

1.1 Digital Signal Processing Hands-On Lab Courses

Nearly all electrical engineering curricula include DSP courses. DSP lab or design courses are also being offered at many universities concurrently or as follow-ups to DSP theory courses. These hands-on lab courses have played a major role in better understanding of DSP concepts. A number of textbooks, e.g. [1–5], have been written to provide the teaching materials for DSP lab courses. The programming language used in these textbooks consists of either C, MATLAB®, or Assembly, which are all text-based languages. In addition to these text-based languages, it is becoming important for students to gain experience in block-based or graphical (G) programming or environment for the purpose of designing DSP systems in a relatively short amount of time. Graphical programming offers an interactive and a more intuitive approach toward building DSP systems. Thus, the main objective of this book is to provide a block-based or system-level programming approach in DSP lab courses. The system-level programming environment chosen is LabVIEW.

Laboratory Virtual Instrumentation Engineering Workbench (LabVIEW) is a graphical programming environment developed by National Instruments (NI) which allows performing high-level or system-level designs. It uses a graphical programming language to create so-called Virtual Instrument (VI) blocks in an intuitive flowchart-like manner. A design is achieved by integrating different blocks, components, or subsystems within a graphical framework. LabVIEW provides data acquisition, analysis, and visualization features well suited for DSP system design. It is also an open environment accommodating MATLAB and C Dynamic Link Libraries (DLLs).

This book is written primarily for those who are already familiar with signal processing concepts and are interested in designing signal processing systems without needing them to be proficient C or MATLAB programmers. After familiarizing the reader with LabVIEW, the book covers a LabVIEW-based approach to generic experiments encountered in a typical DSP lab course. It brings together in one place the information scattered in several NI LabVIEW manuals to provide the necessary tools and know-how for designing signal processing systems within a one-semester lab course. This book can also be used as a self-study LabVIEW guide toward designing and analyzing signal processing systems.

In addition, for those interested in DSP hardware implementation, two chapters in the book are dedicated to executing selected portions of a LabVIEW designed system on an actual DSP processor. The DSP processor chosen is TMS320C6000. This processor has been manufactured by Texas Instruments (TI) for computationally intensive signal processing applications. The DSP hardware utilized to interface with

LabVIEW is the widely adopted TI's C6416 or C6713 DSP Starter Kit (DSK) board. It should be mentioned that since the DSP hardware implementation aspect of the labs (which includes C programs) is independent of the LabVIEW implementation, those who are not interested in the DSP hardware implementation may skip these two chapters.

1.2 Organization

The book includes 12 chapters and 12 labs. After this introduction, the LabVIEW programming environment is presented in Chapter 2. Lab 1 and Lab 2 in Chapter 2 provide a tutorial on getting familiar with the LabVIEW programming environment. Lab 1 provides a general introduction to LabVIEW, and Lab 2 covers building signal processing systems graphically. Lab 2 also shows how to incorporate M-file nodes or blocks within LabVIEW. The topic of analog-to-digital signal conversion is presented in Chapter 3 followed by Lab 3 covering signal sampling experiments. Chapter 4 involves digital filtering. Lab 4 in Chapter 4 shows how to use LabVIEW to design FIR and IIR digital filters. In Chapter 5, fixed-point versus floating-point implementation issues are discussed, followed by Lab 5 covering data type and fixed-point effect experiments. In Chapter 6, the topic of adaptive filtering is discussed. Lab 6 in Chapter 6 covers two adaptive filtering systems consisting of system identification and noise cancellation. Chapter 7 presents frequency domain processing, followed by Lab 7 covering the three widely used transforms in signal processing: fast Fourier transform (FFT), short-time Fourier transform (STFT), and discrete wavelet transform (DWT). Chapter 8 discusses the implementation of a LabVIEW-designed system on the TMS320C6000 DSP processor. First, an overview of the TMS320C6000 architecture is provided. Then, in Lab 8, a tutorial is presented to show how to use the Code Composer Studio (CCStudio) software development tool to achieve the DSP hardware implementation. As a continuation of Chapter 8, Chapter 9 and Lab 9 discuss the issues related to the interfacing of LabVIEW and the DSP processor. Chapters 10 through 12 and Labs 10 through 12, respectively, discuss the following three DSP systems or project examples that are designed in a hybrid mode or a combination of graphical and textual modes: (i) dual tone multi-frequency (DTMF) signaling, (ii) software-defined radio, and (iii) cochlear implant simulator.

1.3 Software Installation

LabVIEW 8.5, which is the latest version at the time of this writing, can be installed by running *setup.exe* on the LabVIEW Core DVD. Some lab portions use the LabVIEW toolkits "Digital Filter Design," "Advanced Signal Processing," and "DSP Test

Integration for TI DSP." The toolkit "Digital Filter Design" appears under the Lab-VIEW Core DVD and can be included while installing LabVIEW 8.5. The toolkits "Advanced Signal Processing" and "DSP Test Integration for TI DSP" appear on the Signal Processing and Communications DVD and can be installed by running *setup.exe* on this DVD. To generate C DLLs, it is required to have Microsoft Visual Studio® or a similar C development environment installed. To use the MATLAB script node feature of LabVIEW, it is required to have MATLAB Version 6.0 or later installed.

If one desires to run parts of a LabVIEW-designed system on a DSP processor, then it is required to install the Code Composer Studio (CCStudio) software tool by running *setup.exe* on the CCStudio CD. In the DSK related labs, CCStudio v3.0 is used.

The accompanying CD includes all the files necessary for running the labs covered throughout the book.

1.4 Updates

Considering that any programming environment goes through enhancements and updates, it is expected that there will be updates of LabVIEW and its toolkits. To accommodate for such updates and to make sure that the labs provided in the book can still be used in DSP lab courses, any new version of the labs will be posted at the website http://www.utdallas.edu/~kehtar/LabVIEW for easy access. It is recommended that this website is periodically checked to download any necessary updates.

1.5 Bibliography

[1] N. Kehtarnavaz, *Real-Time Digital Signal Processing Based on the TMS320C6000*, Elsevier, 2005.

[2] S. Kuo and W-S. Gan, *Digital Signal Processors: Architectures, Implementations, and Applications*, Prentice-Hall, 2005.

[3] R. Chassaing, *DSP Applications Using C and the TMS320C6x DSK*, Wiley Inter-Science, 2002.

[4] T. Welch, C. Wright and M. Morrow, *Real-Time Digital Signal Processing from MATLAB to C with the TMS320C6x DSK*, CRC Press, 2006.

[5] L. Tan, *Digital Signal Processing: Fundamentals and Applications*, Elsevier, 2007.

LabVIEW Graphical Programming Environment

LabVIEW constitutes a graphical programming environment that allows one to design and analyze a DSP system in a shorter time as compared to text-based programming environments. LabVIEW graphical programs are called Virtual Instruments (VIs). VIs run based on the concept of data flow programming. This means that execution of a block or a graphical component is dependent on the flow of data, or more specifically a block executes when data are made available at all of its inputs. Output data of the block are then sent to all other connected blocks. Data flow programming allows multiple operations to be performed in parallel, since its execution is determined by the flow of data and not by sequential lines of code.

2.1 Virtual Instruments (VIs)

A VI consists of two major components, which include a Front Panel (FP) and a Block Diagram (BD). An FP provides the user-interface of a program, whereas a BD incorporates its graphical code. When a VI is located within the block diagram of another VI, it is called a subVI. LabVIEW VIs are modular, meaning that any VI or subVI can be run by itself.

2.1.1 Front Panel and Block Diagram

An FP contains the user interfaces of a VI shown in a BD. Inputs to a VI are represented by controls. Knobs, pushbuttons, and dials are a few examples of controls. Outputs from a VI are represented by indicators. Graphs, LEDs (light indicators), and meters are a few examples of indicators. As a VI runs, its FP provides a display or user interface of controls (inputs) and indicators (outputs).

A BD contains terminal icons, nodes, wires, and structures. Terminal icons are interfaces through which data are exchanged between an FP and a BD. They correspond to controls or indicators that appear on an FP. Whenever a control or indicator is placed on an FP, a terminal icon gets added to the corresponding BD. A node represents an object which has input and/or output connectors and performs a certain function. SubVIs and functions are examples of nodes. Wires establish the flow of data in a BD. Structures are used to control the flow of a program such as repetitions or conditional executions. Figure 2-1 shows what an FP and a BD window look like.

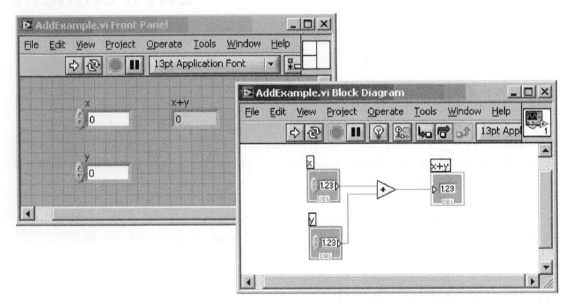

Figure 2-1: LabVIEW windows: Front Panel and Block Diagram.

2.1.2 Icon and Connector Pane

A VI icon is a graphical representation of a VI. It appears in the top right corner of a BD or an FP window. When a VI is inserted in a BD as a subVI, its icon gets displayed.

A connector pane defines inputs (controls) and outputs (indicators) of a VI. The number of inputs and outputs can be changed by using different connector pane patterns. In Figure 2-1, a VI icon is shown at the top right corner of the BD and its corresponding connector pane having two inputs and one output is shown at the top right corner of the FP.

2.2 Graphical Environment

2.2.1 Functions Palette

The Functions palette, shown in Figure 2-2, provides various function VIs or blocks for building a system. This palette can be displayed by right-clicking on an open area of a BD. Note that this palette can be displayed only in a BD.

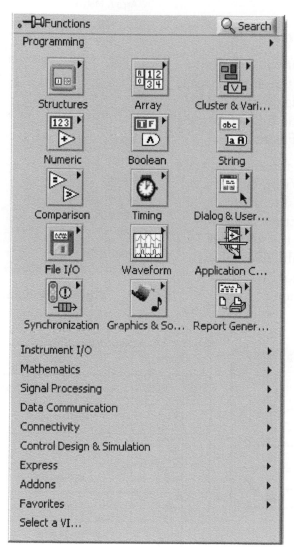

Figure 2-2: Functions palette.

2.2.2 Controls Palette

The Controls palette, shown in Figure 2-3, provides controls and indicators of an FP. This palette can be displayed by right-clicking on an open area of an FP. Note that this palette can be displayed only in an FP.

2.2.3 Tools Palette

The Tools palette provides various operation modes of the mouse cursor for building or debugging a VI. The Tools palette and the frequently used tools are shown in Figure 2-4.

Each tool is utilized for a specific task. For example, the Wiring tool is used to wire objects in a BD. If the automatic tool selection mode is enabled by clicking the **Automatic Tool Selection** button, LabVIEW selects the best matching tool based on a current cursor position.

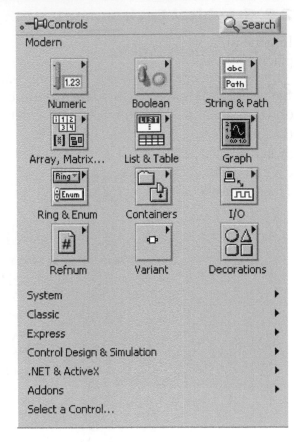

Figure 2-3: Controls palette.

Icon	Tool
	Automatic Tool Selection
	Operating tool
	Positioning tool
	Labeling tool
	Wiring tool
	Probe tool

Figure 2-4: Tools palette.

2.3 Building a Front Panel

In general, a VI is put together by going back and forth between an FP and a BD, placing inputs/outputs on the FP and building blocks on the BD.

2.3.1 Controls

Controls make up the inputs to a VI. Controls grouped in the Numeric Controls palette (**Controls » Express » Num Ctrls**) are used for numerical inputs, grouped in the Buttons & Switches palette (**Controls » Express » Buttons**) for Boolean inputs, and grouped in the Text Controls palette (**Controls » Express » Text Ctrls**) for text and enumeration inputs. These control options are displayed in Figure 2-5.

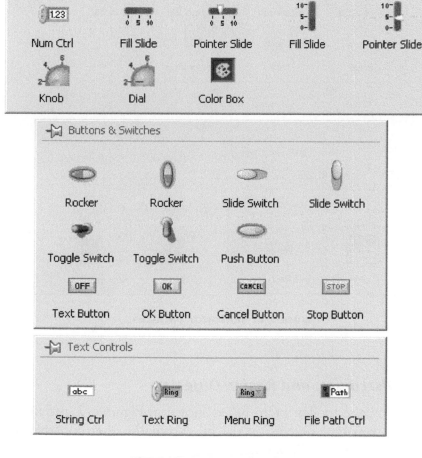

Figure 2-5: Control palettes.

2.3.2 Indicators

Indicators make up the outputs of a VI. Indicators grouped in the Numeric Indicators palette (**Controls » Express » Num Inds**) are used for numerical outputs, grouped in the LEDs palette (**Controls » Express » LEDs**) for Boolean outputs, grouped in the Text Indicators palette (**Controls » Express » Text Inds**) for text outputs, and grouped in the Graph Indicators palette (**Controls » Express » Graph Indicators**) for graphical outputs. These indicator options are displayed in Figure 2-6.

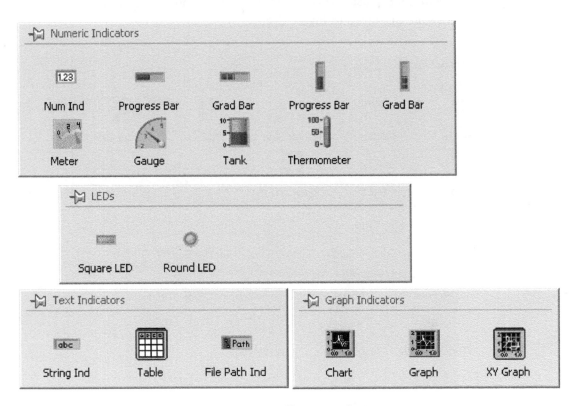

Figure 2-6: Indicator palettes.

2.3.3 Align, Distribute, and Resize Objects

The menu items on the toolbar of an FP, as shown in Figure 2-7, provide options to align and orderly distribute objects on the FP. Normally, after controls and indicators are placed on an FP, one uses these options to tidy up their appearance.

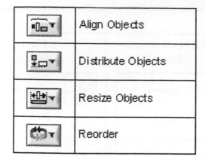

Figure 2-7: Menu for align, distribute, resize, and reorder objects.

2.4 Building a Block Diagram

2.4.1 Express VI and Function

Express VIs denote higher-level VIs that have been configured to incorporate lower-level VIs or functions. These VIs are displayed as expandable nodes with a blue background. Placing an Express VI in a BD brings up a configuration window allowing adjustment of its parameters. As a result, Express VIs demand less wiring. A configuration window can be brought up by double-clicking on its Express VI.

Basic operations such as addition or subtraction are represented by functions. Figure 2-8 shows three examples corresponding to three types of a BD object (VI, Express VI, and function).

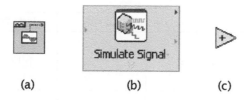

(a)   (b)   (c)

Figure 2-8: Block Diagram objects (a) VI, (b) Express VI, and (c) function.

A subVI or an Express VI can be displayed as icons or expandable nodes. If a subVI is displayed as an expandable node, the background appears yellow. Icons are used to save space in a BD, while expandable nodes are used to provide easier wiring or better readability. Expandable nodes can be resized to show their connection nodes more clearly. Three appearances of a VI/Express VI are shown in Figure 2-9.

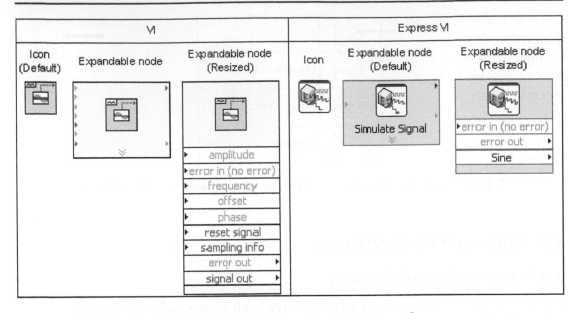

Figure 2-9: Icon versus expandable node.

2.4.2 Terminal Icons

FP objects get displayed as terminal icons in a BD. A terminal icon exhibits an input or output as well as its data type. Figure 2-10 shows two terminal icon examples consisting of a double precision numerical control and indicator. As shown in this figure, terminal icons can be displayed as data type terminal icons to conserve space in a BD.

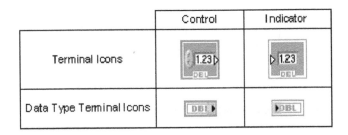

Figure 2-10: Terminal icon examples displayed in a BD.

2.4.3 Wires

Wires transfer data from one node to another in a BD. Based on the data type of a data source, the color and thickness of its connecting wires change.

Wires for the basic data types used in LabVIEW are shown in Figure 2-11. Other than the data types shown in this figure, there are some other specific data types. For example, the dynamic data type is always used for Express VIs, and the waveform data type, which corresponds to the output from a waveform generation VI, is a special cluster of components incorporating trigger time, time interval, and data value.

Wire Type	Scalar	1D Array	2D Array	Color
Numeric				Orange (Floating point) Blue (Integer)
Boolean				Green
String				Pink

Figure 2-11: Basic types of wires [2].

2.4.4 Structures

A structure is represented by a graphical enclosure. The graphical code enclosed by a structure gets repeated or executed conditionally. A loop structure is equivalent to a For Loop or a While Loop statement encountered in text-based programming languages, whereas a Case structure is equivalent to an if-else statement.

2.4.4.1 For Loop

A For Loop structure is used to perform repetitions. As illustrated in Figure 2-12, the displayed border indicates a For Loop structure, where the count terminal [N] represents the number of times the loop is to be repeated. It is set by wiring a value from outside the loop to it. The iteration terminal [i] denotes the number of completed iterations, which always starts at zero.

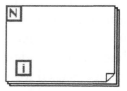

Figure 2-12: For Loop.

2.4.4.2 While Loop

A While Loop structure allows repetitions depending on a condition; see Figure 2-13. The conditional terminal ◉ initiates a stop if the condition is true. Similar to a For Loop, the iteration terminal [i] provides the number of completed iterations, always starting at zero.

Figure 2-13: While Loop.

2.4.4.3 Case Structure

A Case structure, shown in Figure 2-14, allows running different sets of operations depending on the value it receives through its selector terminal, which is indicated by ⍰. In addition to Boolean type, the input to a selector terminal can be of integer, string, or enumerated type. This input determines which case to execute. The case selector ◀ True ▼▶ shows the status being executed. Cases can be added or deleted as needed.

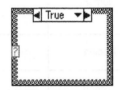

**Figure 2-14:
Case structure.**

2.5 MathScript

MathScript is a new feature of the latest version of LabVIEW (LabVIEW 8.0+) which allows one to perform textual programming in conjunction with graphical programming [6]. It includes various built-in functions and uses matrices and arrays as fundamental data types with built-in operators for data manipulation. User-defined functions can also be added to it. MathScript is compatible with the M-file script syntax that is encountered in MATLAB codes. MathScript possesses an interactive and a programming interface named MathScript Interactive Window and MathScript Node, respectively.

A MathScript Interactive Window is shown in Figure 2-15. Three interfaces— command window, output window, and MathScript window—are shown in this figure. The command window interface is used to enter commands and for script debugging or to view help statements for built-in functions. The output window interface is used for viewing output values. The MathScript window interface is used to display variables, edit scripts, and display command history. Script editing allows the execution of a group of commands.

A MathScript Node represents the textual code via a blue rectangle, as shown in Figure 2-16. Its inputs and outputs are defined on the border of this rectangle

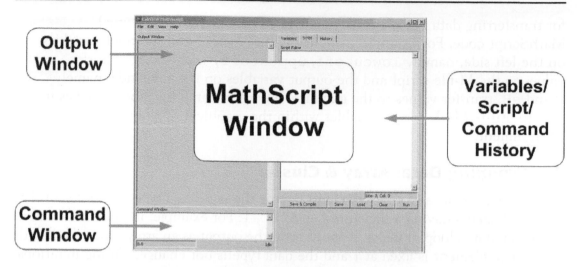

Figure 2-15: MathScript Interactive Window.

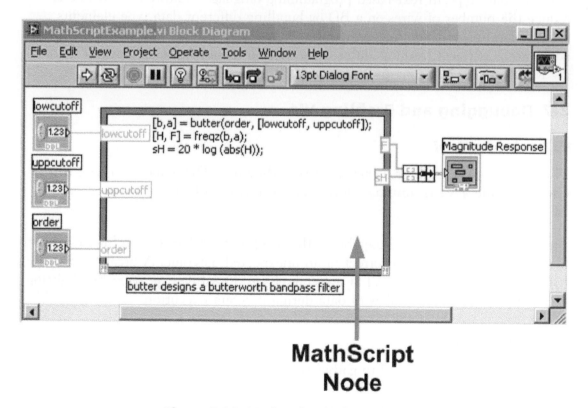

Figure 2-16: MathScript Node Interface.

for transferring data between the graphical environment and a textual MathScript code. For example, as indicated in Figure 2-16, the input variables on the left side, namely `lowcutoff`, `uppcutoff`, and `order`, transfer values to the M-file script and the output variables on the right side, namely F and sH, transfer values to the graphical environment. This process makes it easy to utilize M-file script variables within the graphical programming environment.

2.6 Grouping Data: Array & Cluster

An array represents a group of elements having the same data type. An array consists of data elements having a dimension up to $2^{31}-1$. For example, if a random number is generated in a loop, it makes sense to build the output as an array, since the length of the data element is fixed at 1 and the data type is not changed during iterations.

A cluster consists of a collection of different data type elements, similar to the structure data type in text-based programming languages. Clusters allow one to reduce the number of wires on a BD by bundling different data type elements together and passing them to only one terminal. An individual element can be added to or extracted from a cluster by using the cluster functions such as `Bundle by Name` and `Unbundle by Name`.

2.7 Debugging and Profiling VIs

2.7.1 Probe Tool

The Probe tool is used for debugging VIs as they run. The value on a wire can be checked while a VI is running. Note that the Probe tool can be accessed only in a BD window.

The Probe tool can be used together with breakpoints and execution highlighting to identify the source of an incorrect or an unexpected outcome. A breakpoint is used to pause the execution of a VI at a specific location, while execution highlighting helps one to visualize the flow of data during program execution.

2.7.2 Profile Tool

The Profile tool can be used to gather timing and memory usage information, i.e., how long a VI takes to run and how much memory it consumes. It is necessary to make sure a VI is stopped before setting up a Profile window.

An effective way to become familiar with LabVIEW programming is by going through hands-on examples. Thus, in the two labs that follow in this chapter, most of the key programming features of LabVIEW are learned via building some simple VIs. More detailed information on LabVIEW programming can be found in [1-6].

2.8 Bibliography

[1] National Instruments, *LabVIEW Getting Started with LabVIEW*, Part Number 323427A-01, 2003.

[2] National Instruments, *LabVIEW User Manual*, Part Number 320999E-01, 2003.

[3] National Instruments, *LabVIEW Performance and Memory Management*, Part Number 342078A-01, 2003.

[4] National Instruments, *Introduction to LabVIEW Six-Hour Course*, Part Number 323669B-01, 2003.

[5] Robert H. Bishop, *Learning with LabVIEW 7 Express*, Prentice-Hall, 2003.

[6] National Instruments, *Inside LabVIEW MathScript*, http://zone.ni.com/devzone/conceptd.nsf/webmain

Lab 1: Getting Familiar with LabVIEW: Part I

The objective of this first lab is to provide an initial hands-on experience in building a VI. For detailed explanations of the LabVIEW features mentioned here, the reader is referred to [1]. LabVIEW 8.5 can get launched by double-clicking on the LabVIEW 8.5 icon. The dialog box shown in Figure L1-1 should appear.

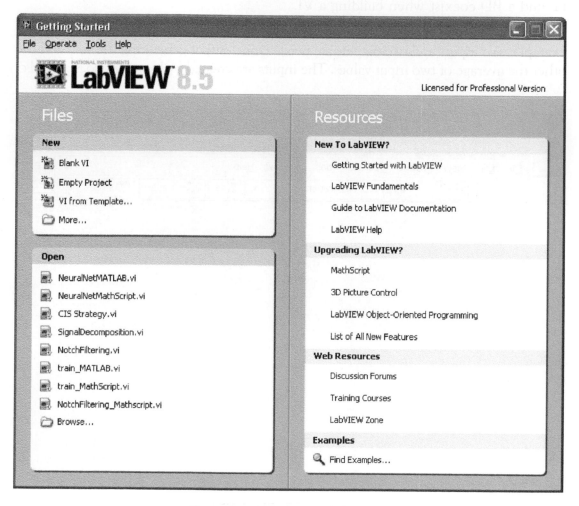

Figure L1-1: Starting LabVIEW.

L1.1 Building a Simple VI

To become familiar with the LabVIEW programming environment, it is found to be more effective if one starts by going through a simple example. The example presented here consists of calculating the sum and average of two input values. This example is described in a step-by-step fashion in the following sections.

L1.1.1 VI Creation

To create a new VI, click on the **Blank VI** under New; see Figure L1-1. This step can also be done by choosing **File » New VI** from the menu. As a result, a blank FP and a blank BD window appear, as shown in Figure L1-2. It should be remembered that an FP and a BD coexist when building a VI.

Clearly, the number of inputs and outputs to a VI is dependent on its function. In this example, two inputs and two outputs are needed, one output generating the sum and the other the average of two input values. The inputs are created by locating two Numeric

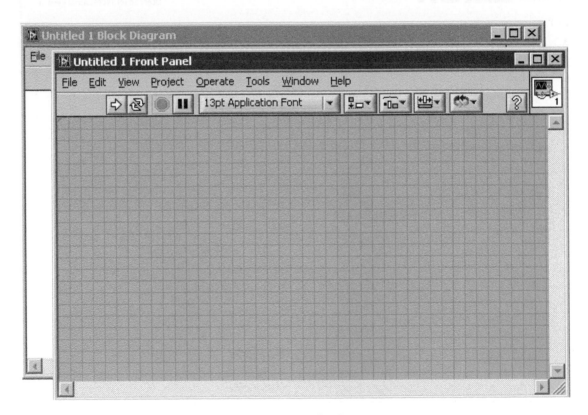

Figure L1-2: Blank VI.

Controls on the FP. This is done by right-clicking on an open area of the FP to bring up the Controls palette, followed by choosing **Controls » Modern » Numeric » Numeric Control**. Each numeric control automatically places a corresponding terminal icon on the BD. Double-clicking on a numeric control highlights its counterpart on the BD, and vice versa.

Next, let us label the two inputs as x and y. This is achieved by using the Labeling tool from the **Tools** palette, which can be displayed by choosing **View » Tools Palette** from the menu bar. Choose the Labeling tool and click on the default labels, Numeric and Numeric 2, in order to edit them. Alternatively, if the automatic tool selection mode is enabled by clicking **Automatic Tool Selection** in the **Tools** palette, the labels can be edited by simply double-clicking on the default labels. Editing a label on the FP changes its corresponding terminal icon label on the BD, and vice versa.

Similarly, the outputs are created by locating two Numeric Indicators (**Controls » Modern » Numeric » Numeric Indicator**) on the FP. Each numeric indicator automatically places a corresponding terminal icon on the BD. Edit the labels of the indicators to read Sum and Average.

For a better visual appearance, objects on an FP window can be aligned, distributed, and resized using the appropriate buttons appearing on the FP toolbar. To do this, select the objects to be aligned or distributed and apply the appropriate option from the toolbar menu. Figure L1-3 shows the configuration of the FP just created.

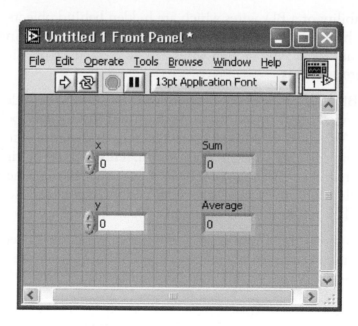

Figure L1-3: FP configuration.

Now, let us build a graphical code on the BD to perform the summation and averaging operations. Note that <Ctrl-E> toggles between an FP and a BD window. If one finds the objects on a BD are too close to insert other functions or VIs in between, a horizontal or vertical space can be inserted by holding down the <Ctrl> key to create space horizontally and/or vertically. As an example, Figure L1-4(b) illustrates a horizontal space inserted between the objects shown in Figure L1-4(a).

Next, place an Add function (**Functions » Express » Arithmetic & Comparison » Express Numeric » Add**) and a Divide function (**Functions » Express » Arithmetic & Comparison » Express Numeric » Divide**) on the BD. The divisor, in our case 2, needs to be entered in a Numeric Constant (**Functions » Express » Arithmetic & Comparison » Express Numeric » Numeric Constant**) and connected to the y terminal of the Divide function using the Wiring tool.

To have a proper data flow, functions, structures, and terminal icons on a BD need to be wired. The Wiring tool is used for this purpose. To wire these objects, point the Wiring tool at a terminal of a function or a subVI to be wired, click on the terminal, drag the mouse to a destination terminal, and click once again. Figure L1-5 illustrates the wires placed between the terminals of the numeric controls and the input terminals of the add function. Notice that the label of a terminal is displayed whenever the cursor is moved over it if the automatic tool selection mode is enabled. Also, note that the Run button ⏵ on the toolbar remains broken until the wiring process is completed.

For better readability of a BD, wires which are hidden behind objects or crossed over other wires can be cleaned up by right-clicking on them and choosing **Clean Up Wire** from the shortcut menu. Any broken wires can be cleared by pressing <Ctrl-B> or **Edit » Remove Broken Wires**.

The label of a BD object, such as a function, can be shown (or hidden) by right-clicking on the object and checking (or unchecking) **Visible Items » Label** from the shortcut menu. Also, a terminal icon corresponding to a numeric control or indicator can be shown as a data type terminal icon. This is done by right-clicking on the terminal icon and unchecking **View As Icon** from the shortcut menu. Figure L1-6 shows an example where the numeric controls and indicators are shown as data type terminal icons. The notation DBL represents double precision data type.

It is worth pointing out that there exists a shortcut to build the preceding VI. Instead of choosing the numeric controls, indicators, or constants from the Controls or Functions palette, one can use the shortcut menu **Create**, activated by right-clicking on a terminal of a BD object such as a function or a subVI. As an example of this approach, create a blank VI and locate an Add function. Right-click on its x terminal and choose **Create » Control** from the shortcut menu to create and wire a

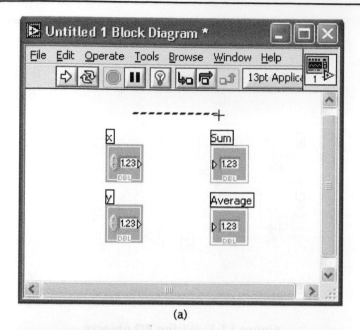

(a)

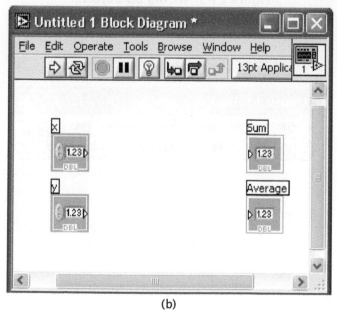

(b)

Figure L1-4: Inserting horizontal/vertical space: (a) creating space while holding down the <Ctrl> key, and (b) inserted horizontal space.

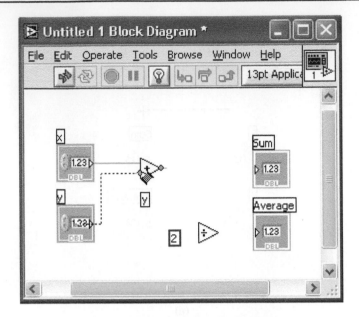

Figure L1-5: Wiring BD objects.

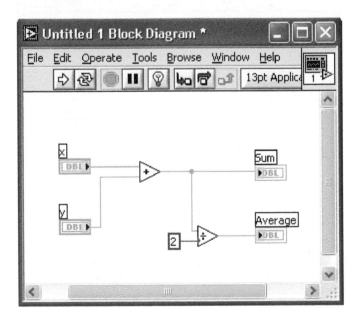

Figure L1-6: Completed BD.

numeric control or input. This locates a numeric control on the FP as well as a corresponding terminal icon on the BD. The label is automatically set to x. Create a second numeric control by right-clicking on the y terminal of the Add function. Next, right-click on the output terminal of the Add function and choose **Create » Indicator** from the shortcut menu. A data type terminal icon, labeled as x+y, is created on the BD as well as a corresponding numeric indicator on the FP.

Next, right-click on the y terminal of the Divide function to choose **Create » Constant** from the shortcut menu. This creates a Numeric Constant as the divisor and wires its y terminal. Type the value 2 in the numeric constant. Right-click on the output terminal of the Divide function, labeled as x/y, and choose **Create » Indicator** from the shortcut menu. In case a wrong option is chosen, the terminal does not get wired. A wrong terminal option can be easily changed by right-clicking on the terminal and choosing **Change to Control** or **Change to Constant** from the shortcut menu.

To save the created VI for later use, choose **File » Save** from the menu or press <Ctrl-S> to bring up a dialog box to enter a name. Type Sum and Average as the VI name and click **Save**.

To test the functionality of the VI, enter some sample values in the numeric controls on the FP and run the VI by choosing **Operate » Run**, by pressing <Ctrl-R>, or by clicking the Run button on the toolbar. From the displayed output values in the numeric indicators, the functionality of the VI can be verified. Figure L1-7 illustrates the outcome after running the VI with two inputs 10 and 30.

L1.1.2 SubVI Creation

If a VI is to be used as part of a higher level VI, its connector pane needs to be configured. A connector pane assigns inputs and outputs of a subVI to its terminals through which data are exchanged. A connector pane can be

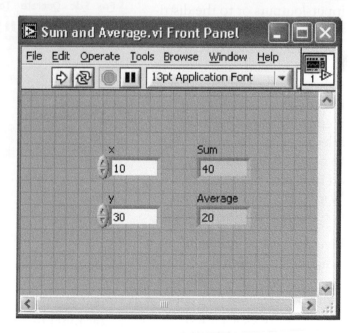

Figure L1-7: VI verification.

displayed by right-clicking on the top right corner icon of an FP and selecting **Show Connector** from the shortcut menu.

The default pattern of a connector pane is determined based on the number of controls and indicators. In general, the terminals on the left side of a connector pane pattern are used for inputs, and the ones on the right side for outputs. Terminals can be added to or removed from a connector pane by right-clicking and choosing **Add Terminal** or **Remove Terminal** from the shortcut menu. If a change is to be made to the number of inputs/outputs or to the distribution of terminals, a connector pane pattern can be replaced with a new one by right-clicking and choosing **Patterns** from the shortcut menu. Once a pattern is selected, each terminal needs to be reassigned to a control or an indicator by using the Wiring tool, or by enabling the automatic tool selection mode.

Figure L1-8(a) illustrates assigning a terminal of the Sum and Average VI to a numeric control. The completed connector pane is shown in Figure L1-8(b).

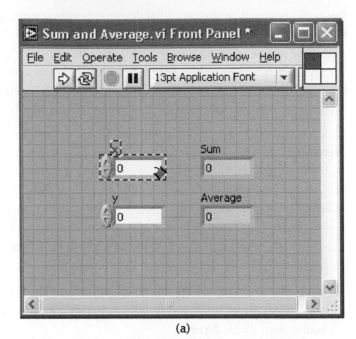

(a)

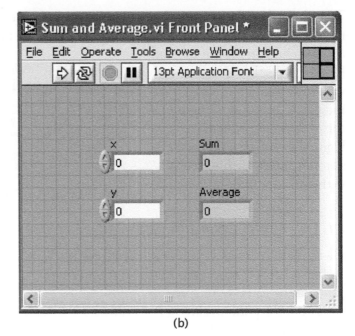

(b)

Figure L1-8 Connector pane: (a) assigning a terminal to a control, and (b) terminal assignment completed.

Notice that the output terminals have thicker borders. The color of a terminal reflects its data type.

Considering that a subVI icon is displayed on the BD of a higher level VI, it is important to edit the subVI icon for it to be explicitly identified. Double-clicking on the top right corner icon of a BD brings up the Icon Editor. The tools provided in the Icon Editor are very similar to those encountered in other graphical editors, such as Microsoft Paint. An editing of the icon for the Sum and Average VI is illustrated in Figure L1-9.

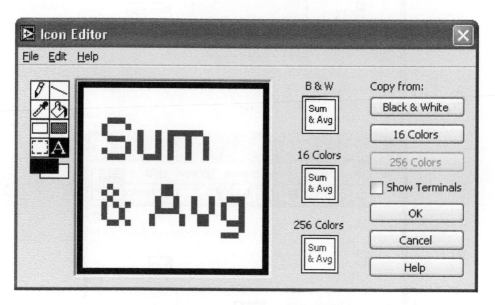

Figure L1-9: Editing subVI icon.

A subVI can also be created from a section of a VI. To do so, select the nodes on the BD to be included in the subVI, as shown in Figure L1-10(a). Then, choose **Edit »
Create SubVI**. This inserts a new subVI icon. Figure L1-10(b) illustrates the BD with an inserted subVI. This subVI can be opened and edited by double-clicking on its icon on the BD. Save this subVI as *Sum* and *Average.vi*. This subVI performs the same function as the original Sum and Average VI.

In Figure L1-11, the completed FP and BD of the Sum and Average VI are shown.

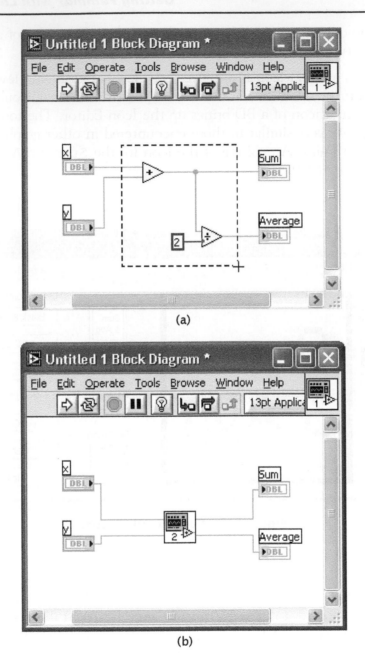

(a)

(b)

Figure L1-10 Creating a subVI: (a) selecting nodes to make a subVI, and (b) inserted subVI icon.

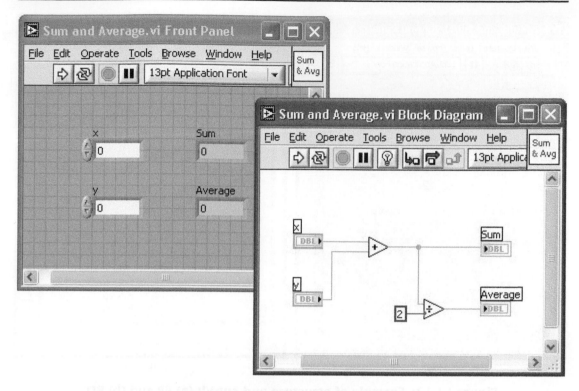

Figure L1-11: Sum and Average VI.

L1.2 Using Structures and SubVIs

Let us now consider another example to demonstrate the use of structures and subVIs. In this example, a VI is used to show the sum and average of two input values in a continuous fashion. The two inputs can be altered by the user. If the average of the two inputs becomes greater than a preset threshold value, an LED warning light is lit.

As the first step to build such a VI, build an FP as shown in Figure L1-12(a). For the inputs, consider two Knobs (**Controls » Modern » Numeric » Knob**). Adjust the size of the knobs by using the Positioning tool. Properties of knobs such as precision and data type can be modified by right-clicking and choosing **Properties** from the shortcut menu. A Knob Properties dialog box is brought up, and an **Appearance** tab is shown by default. Edit the label of one of the knobs to read Input 1. Select the **Data Range** tab, and click **Representation** to change the data type from double precision to byte by selecting **Byte** among the displayed data types. This can also be achieved by right-clicking on the knob and choosing **Representation » Byte** from the shortcut menu. In the **Data Range** tab, a default value needs to be specified.

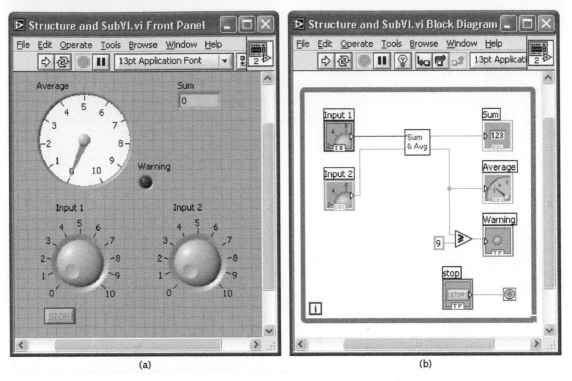

Figure L1-12: Example of structure and subVI: (a) FP and (b) BD.

In this example, the default value is considered to be 0. The default value can be set by right-clicking on the control and choosing **Data Operations » Make Current Value Default** from the shortcut menu. Also, this control can be set to a default value by right-clicking and choosing **Data Operations » Reinitialize to Default Value** from the shortcut menu.

Label the second knob as Input 2 and repeat all the adjustments as done for the first knob except for the data representation part. The data type of the second knob is specified to be double precision in order to demonstrate the difference in the outcome. As the final step of configuring the FP, align and distribute the objects using the appropriate buttons on the FP toolbar.

To set the outputs, locate and place a Numeric Indicator, a Round LED (**Controls » Modern » Boolean » Round LED**), and a Gauge (**Controls » Modern » Numeric » Gauge**). Edit the labels of the indicators as shown in Figure L1-12(a).

Now let us build the BD as shown in Figure L-12(b). There are five control and indicator icons already appearing on the BD. Right-click on an open area of the BD to bring up the **Functions** palette and then choose **Select a VI...**. This brings up a file dialog box. Navigate to the Sum and Average VI in order to place it on the BD. This subVI is displayed as an icon on the BD. Wire the numeric controls, Input 1 and Input 2, to the x and y terminals, respectively. Also, wire the Sum terminal of the subVI to the numeric indicator labeled Sum and the Average terminal to the gauge indicator labeled Average.

A Greater or Equal? function is located from **Functions » Programming » Comparison » Greater or Equal?** in order to compare the average output of the subVI with a threshold value. Create a wire branch on the wire between the Average terminal of the subVI and its indicator via the Wiring tool. Then, extend this wire to the x terminal of the Greater or Equal? function. Right-click on the y terminal of the Greater or Equal? function and choose **Create » Constant** in order to place a Numeric Constant. Enter 9 in the numeric constant. Then, wire the Round LED, labeled as Warning, to the x>=y? terminal of this function to provide a Boolean value.

In order to run the VI continuously, one uses a While Loop structure. Choose **Functions » Programming » Structures » While Loop** to create a While Loop. Change the size by dragging the mouse to enclose the objects in the While Loop as illustrated in Figure L1-13.

Once this structure is created, its boundary together with the loop iteration terminal ⒤, and conditional terminal ◉ get shown on the BD. If the While Loop is created by using **Functions » Programming » Structures » While Loop**, then the Stop Button is not included as part of the

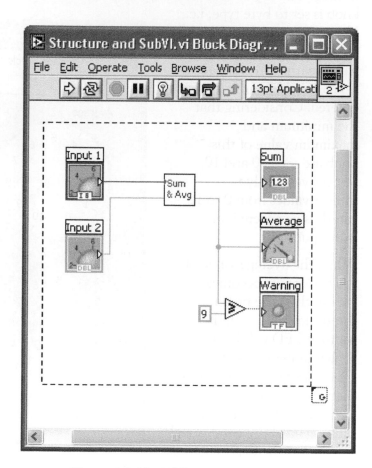

Figure L1-13: While Loop enclosure.

structure. This button can be created by right-clicking on the conditional terminal and choosing **Create » Control** from the shortcut menu. A Boolean condition can be wired to a conditional terminal, instead of a stop button, in order to stop the loop programmatically.

As the final step, tidy up the wires, nodes, and terminals on the BD using the **Align object** and **Distribute object** options on the BD toolbar. Then, save the VI in a file named *Structure and SubVI.vi*.

Now run the VI to verify its functionality. After clicking the Run button on the toolbar, adjust the knobs to alter the inputs. Verify whether the average and sum are displayed correctly in the gauge and numeric indicators. Note that only integer values can be entered via the Input 1 knob, whereas real values can be entered via the Input 2 knob. This is due to the data types associated with these knobs. The Input 1 knob is set to byte type, i.e., I8 or 8-bit signed integer. As a result, only integer values within the range −128 and 127 can be entered. Considering that the minimum and maximum value of this knob are set to 0 and 10, respectively, only integer values from 0 to 10 can thus be entered for this input.

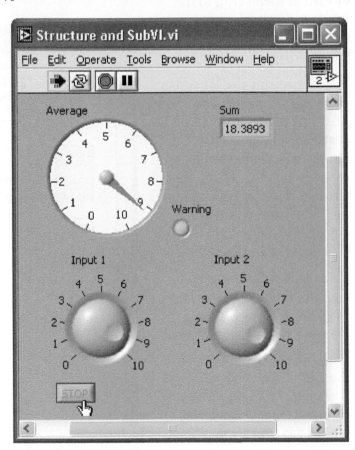

When the average value of the two inputs becomes greater than the preset threshold value of 9, the warning LED will light up, as shown in Figure L-14. Click the stop button on the FP to stop the VI. Otherwise, the VI keeps running until the conditional terminal of the While Loop becomes true.

Figure L1-14: FP as VI runs.

L1.3 Create an Array with Indexing

Auto-indexing enables one to read/write each element from/to a data array in a loop structure. In this section, this feature is covered.

Let us first locate a `For Loop` (**Functions » Programming » Structures » For Loop**). Right-click on its count terminal and choose **Create » Constant** from the shortcut menu to set the number of iterations. Enter 10 so that the code inside it gets repeated 10 times. Note that the current loop iteration count, which is read from the iteration terminal, starts at index 0 and ends at index 9.

Place a `Random Number (0-1)` function (**Functions » Programming » Numeric » Random Number (0–1)**) inside the `For Loop` and wire the output terminal of this function, `number (0 to 1)`, to the border of the `For Loop` to create an output tunnel. The tunnel appears as a box with the array symbol [ ] inside it. For a `For Loop`, auto-indexing is enabled by default, whereas for a `While Loop`, it is disabled by default. Create an indicator on the tunnel by right-clicking and choosing **Create » Indicator** from the shortcut menu. This creates an array indicator icon outside the loop structure on the BD. Its wire appears thicker due to its array data type. Also, another indicator representing the array index gets displayed on the FP. This indicator is of array data type and can be resized as desired. In this example, the size of the array is specified as 10 to display all the values, considering that the number of iterations of the `For Loop` is set to be 10.

Create a second output tunnel by wiring the output of the `Random Number (0-1)` function to the border of the loop structure; then right-click on the tunnel and choose **Disable indexing** from the shortcut menu to disable auto-indexing. When one does this, the tunnel becomes a filled box representing a scalar value. Create an indicator on the tunnel by right-clicking and choosing **Create » Indicator** from the shortcut menu. This sets up an indicator of scalar data type outside the loop structure on the BD.

Next, create a third indicator on the `Number (0 to 1)` terminal of the `Random Number (0-1)` function located in the `For Loop` to observe the values coming out. To do this, right-click on the output terminal or on the wire connected to this terminal and choose **Create » Indicator** from the shortcut menu.

Place a `Time Delay` Express VI (**Functions » Programming » Timing » Time Delay**) to delay the execution in order to have enough time to observe a current value. A configuration window is brought up for specifying the delay time in seconds. Enter the value 0.1 to wait 0.1 seconds at each iteration. Note that the `Time Delay` Express VI is shown as an icon in Figure L1-15 in order to have a more compact display.

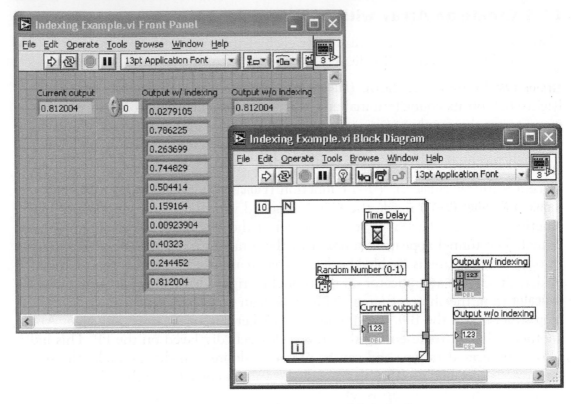

Figure L1-15: Creating array with indexing.

Save the VI as *Indexing Example.vi* and run it to observe its functionality. From the output displayed on the FP, a new random number should get displayed every 0.1 second on the indicator residing inside the loop structure. However, no data will be available from the indicators outside the loop structure until the loop iterations end. An array of 10 elements should be generated from the indexing-enabled tunnel, while only one output, the last element of the array, should be passed from the indexing-disabled tunnel.

L1.4 Debugging VIs: Probe Tool

The Probe tool is used to observe data that are being passed while a VI is running. A probe can be placed on a wire by using the Probe tool or by right-clicking on a wire and choosing **Probe** from the shortcut menu. Probes can also be placed while a VI is running.

Placing probes on wires creates probe windows through which intermediate values can be observed. A probe window can be customized. For example, showing data of array data type via a graph makes debugging easier. To do this, right-click on the wire where an array is being passed and choose **Custom Probe » Controls » Modern » Graph » Waveform Graph** from the shortcut menu.

As an example of using custom probes, a `Waveform Chart` is used here to track the scalar values at probe location 1, a waveform graph to monitor the array at probe location 2, and a regular probe window at probe location 3 to see the values of the `Indexing Example` VI. These probes and their locations are illustrated in Figure L1-16.

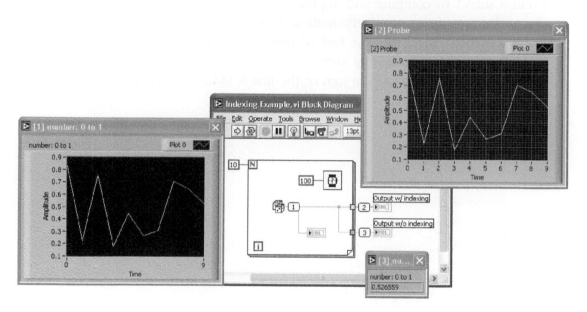

Figure L1-16: Probe tool.

L1.5 Bibliography

[1] National Instruments, *LabVIEW User Manual*, Part Number 320999E-01, 2003.

L1.6 Lab Experiments

Perform the following experiments with and without using the MathScript feature of LabVIEW 8.

1. Build a subVI to compute the product, sum, and difference of two given square matrices A and B.
2. Build a subVI to compute and display the roots of a quadratic equation $ax^2 + bx + c$ for given coefficients a, b, and c.
3. Build a subVI to generate the first 20 numbers of the Fibonacci sequence and store them using an indexing array.
4. Build a subVI to compute the sum of the first n natural numbers for a specified value of n.

Lab 2: Getting Familiar with LabVIEW: Part II

Now that an initial familiarity with the LabVIEW programming environment has been acquired in Lab 1, this second lab shows how a DSP system can be built in LabVIEW. In addition, the hybrid programming approach is introduced.

L2.1 Express VIs Versus Regular VIs

A simple DSP system consisting of signal generation and amplification is covered here. The shape of the input signal (sine, square, triangle, or saw tooth) as well as its frequency and gain are altered by using appropriate FP controls. The system is built with Express VIs first; then the same system is built with regular VIs. This is done in order to illustrate the use of Express VIs versus regular VIs for building a system.

L2.1.1 Building a System VI with Express VIs

The use of Express VIs allows less wiring on a BD. Also, it provides an interactive user interface by which parameter values can be adjusted on the fly. The BD of the signal generation system using Express VIs is shown in Figure L2-1.

To build this BD, locate the `Simulate Signal` Express VI (**Functions » Express » Input » Simulate Signal**) to generate a signal source. This brings up a configuration window, as shown in Figure L2-2. Different types of signals including sine, square, triangle, sawtooth, or DC can be generated with this VI. Enter and adjust the parameters as indicated in Figure L2-2 to simulate a sinewave having a frequency of 200 Hz and an amplitude swinging between –100 and 100. Set the sampling frequency to 8000 Hz. A total of 128 samples spanning a time duration of 15.875 milliseconds (ms) is generated. Note that when the parameters are changed, the modified signal gets displayed instantly in the **Result Preview** graph window.

Next, place a `Scaling and Mapping` Express VI (**Functions » Express » Arithmetic & Comparison » Scaling and Mapping**) to amplify or scale this simulated input signal.

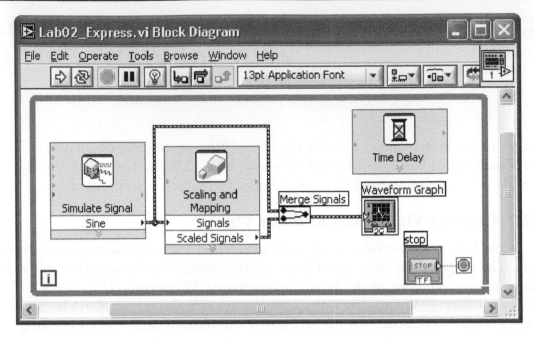

Figure L2-1: BD of signal generation and amplification system using Express VIs [1].

When its configuration dialog box is brought up, as shown in Figure L2-3, choose **Linear (Y=mx+b)** and enter 5 in **Slope (m)** to scale the signal 5 times.

Wire the Sine terminal of the Simulate Signal Express VI to the Signals terminal of the Scaling and Mapping Express VI. Note that a wire having a dynamic data type gets created.

To display the output signal, place a waveform graph (**Controls » Modern » Graph » Waveform Graph**) on the FP. The waveform graph can also be created by right-clicking on the Scaled Signals terminal and choosing **Create » Graph Indicator** from the shortcut menu.

Now, in order to observe the original and the scaled signal together in the same graph, wire the Sine terminal of the Simulate Signal Express VI to the waveform graph. This inserts a Merge Signals function on the wire automatically. An automatic insertion of the Merge Signals function occurs when a signal having a dynamic data type is wired to other signals having the same or other data

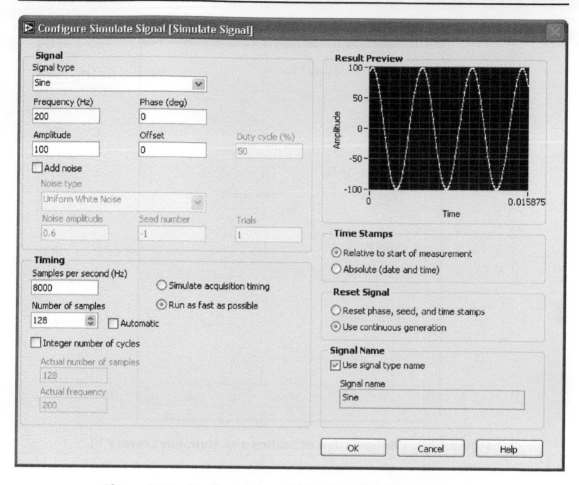

Figure L2-2: Configuration of Simulate Signal Express VI.

types. The `Merge Signals` function combines multiple inputs, thus allowing two signals, consisting of the original and scaled signals, to be handled by one wire. Since both the original and scaled signals are displayed in the same graph, resize the plot legend to display the two labels and markers. The use of the dynamic data type sets the signal labels automatically.

To run the VI continuously, place a `While Loop`. Position the `While Loop` to enclose all the Express VIs and the graph. Now the VI is ready to be run.

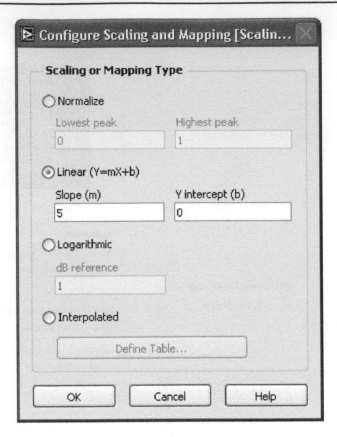

Figure L2-3: Configuration of Scaling and Mapping Express VI.

Run the VI and observe the waveform graph. The output should appear as shown in Figure L2-4. To extend the plot to the right-end of the plotting area, right-click on the waveform graph, choose **X Scale**, and then uncheck **Loose Fit** from the shortcut menu. The graph shown in Figure L2-5 should appear.

If the plot runs too fast, a delay can be placed in the While Loop. To do this, place a Time Delay Express VI (**Functions » Programming » Timing » Time Delay**) and set the delay time to 0.2 in the configuration window. This way, the loop execution is delayed by 0.2 second in the BD appearing in Figure L2-1.

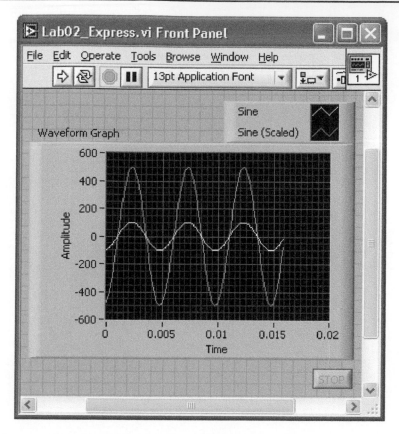

Figure L2-4: FP of signal generation and amplification system.

Although this system runs successfully, no control of the signal frequency and gain is available during its execution, since all the parameters are set in the configuration dialogs of the Express VIs. To gain such a flexibility, one needs to make some modifications.

To change the frequency at run time, place a Vertical Pointer Slide control (**Controls » Modern » Numeric » Vertical Pointer Slide**) on the FP and wire it to the Frequency terminal of the Simulate Signal Express VI. The control is labeled as Frequency. The Express VI can be resized to show more terminals at the

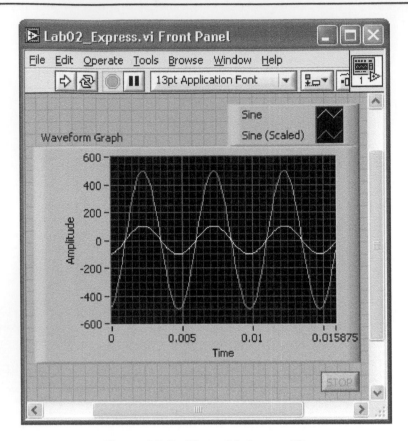

Figure L2-5: Plot with Loose Fit.

bottom of the expandable node. Resize the VI to show an additional terminal below the Sine terminal. Then, click on this new terminal, error out by default, to select Frequency from the list of the displayed terminals.

Next, replace the Scaling and Mapping Express VI with a Multiply function (**Functions » Programming » Numeric » Multiply**). Place another Vertical Pointer Slide control and wire it to the y terminal of the Multiply function to adjust the gain. This control is labeled as Gain. These modifications are illustrated in Figure L2-6.

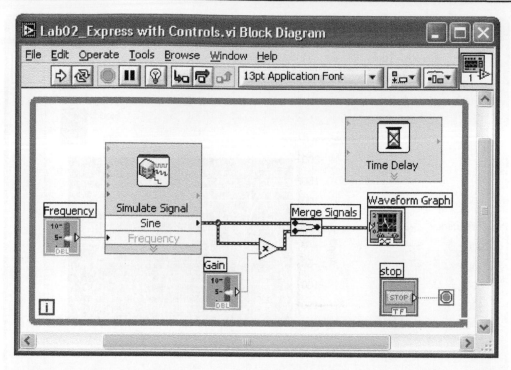

Figure L2-6: BD of signal generation and amplification system with controls.

Now on the FP, set the maximum range of each slide control to 1000 for the Frequency control and 5 for the Gain control, respectively. Also, set the default values for these controls to 200 and 2, respectively.

By running this modified VI, one can observe that the two signals get displayed with the same label, since the source of these signals, i.e., the Sine terminal of the Simulate Signal Express VI, is the same. Also, due to the autoscale feature of the waveform graph, the scaled signal appears unchanged, whereas the Y axis of the waveform graph changes appropriately. This is illustrated in Figure L2-7.

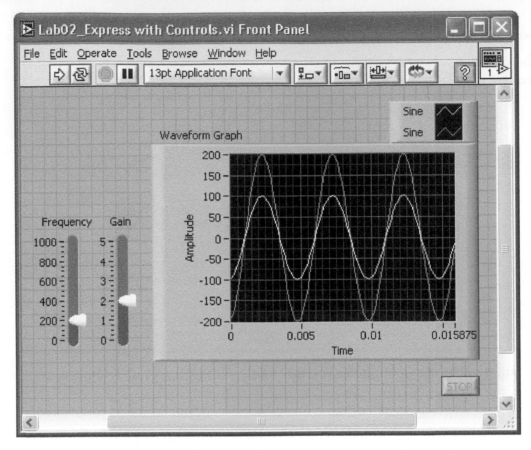

Figure L2-7: Autoscaled graph of two signals shown together.

Let us now modify the property of the waveform graph. In order to disable the autoscale feature, right-click on the waveform graph and uncheck **Y Axis » AutoScale Y**. The maximum and minimum scale can also be adjusted. In this example –600 and 600 are used as the minimum and maximum values, respectively. This is done by modifying the maximum and minimum scale values of the Y axis with the Labeling tool. If the automatic tool selection mode is enabled, just click on the maximum or minimum scale of the Y axis to enter any desired scale value. To modify the labels displayed in the plot legend, right-click and choose **Ignore Attributes**. Then, edit the labels to read Original and Scaled using the Labeling tool. The properties of the waveform graph can also be changed by using its properties dialog box. This box is brought up by right-clicking on the waveform graph and choosing **Properties** from the shortcut menu.

The completed FP is shown in Figure L2-8. With this version of the VI, the frequency of the input signal and the gain of the output signal can be controlled using the controls on the FP.

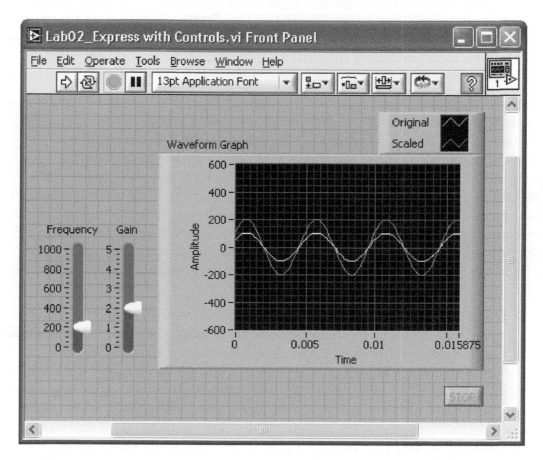

Figure L2-8: FP of signal generation and amplification system with controls.

L2.1.2 Building a System with Regular VIs

In this section, the implementation of the same system discussed in the preceding section is achieved by using regular VIs.

After creating a blank VI, place a `While Loop` (**Functions » Programming » Structures » While Loop**) on the BD, which may need to be resized later. To provide the signal source of the system, place a `Basic Function Generator` VI (**Functions » Programming » Waveform » Analog Waveform » Waveform Generation » Basic Function Generator**) inside the `While Loop`. To configure the parameters of the signal, one needs to wire appropriate controls and constants. To create a control for the signal type, right-click on the `signal type` terminal of the `Basic Function Generator` VI and choose **Create » Control** from the shortcut menu. Note that an enumerated (Enum) type control for the signal gets located on the FP. Four items including sine, triangle, square, and sawtooth are listed in this control.

Next, right-click on the `amplitude` terminal and choose **Create » Constant** from the shortcut menu to create an amplitude constant. Enter 100 in the numeric constant box to set the amplitude of the signal. In order to configure the sampling frequency and the number of samples, create a constant on the `sampling information` terminal by right-clicking and choosing **Create » Constant** from the shortcut menu. This creates a cluster constant which includes two numeric constants. The first element of the cluster shown in the upper box represents the sampling frequency, and the second element shown in the lower box represents the number of samples. Enter 8000 for the sampling frequency and 128 for the number of samples. Note that the same parameters were used in the preceding section.

Now, toggle to the FP by pressing <Ctrl-E> and place two `Vertical Pointer Slide` controls on the FP by choosing **Controls » Modern » Numeric » Vertical Pointer Slide**. Rename the controls `Frequency` and `Gain,` respectively. Set the maximum scale values to 1000 for the `Frequency` control and 5 for the `Gain` control. The `Vertical Pointer Slide` controls create corresponding icons on the BD. Make sure that the icons are located inside the `While Loop`. If not, select the icons and drag them inside the `While Loop`. The `Frequency` control should be wired to the `frequency` terminal of the `Basic Function Generator` VI in order to be able to adjust the frequency at run time. The `Gain` control is used at a later stage.

The output of the Basic Function Generator VI appears in the waveform data type. The waveform data type is a special cluster which bundles three components (t0, dt, and Y) together. The component t0 represents the trigger time of the waveform; dt, the time interval between two samples; and Y, data values of the waveform.

Next, the generated signal needs to be scaled based on a gain factor. This is done by using a Multiply function (**Functions » Programming » Numeric » Multiply**) and a second Vertical Pointer Slide control, named Gain. Wire the generated waveform out of the signal out terminal of the Basic Function Generator VI to the x terminal of the Multiply function. Also, wire the Gain control to the y terminal of the Multiply function.

Recall that the Merge Signals function is used to combine two signals having dynamic data types into the same wire. To achieve the same outcome with regular VIs, place a Build Array function (**Functions » Programming » Array » Build Array**) to build a 2D array, i.e., two rows (or columns) of one dimensional signal. Resize the Build Array function to have two input terminals. Wire the original signal to the upper terminal of the Build Array function and the output of the Multiply function to the lower terminal. Remember that the Build Array function is used to concatenate arrays or build *n*-dimensional arrays. Since the Build Array function is used for comparing the two signals, make sure that the **Concatenate Inputs** option is unchecked from the shortcut menu. More details on the use of the Build Array function can be found in [2].

A waveform graph (**Controls » Modern » Graph » Waveform Graph**) is then placed on the FP. Wire the output of the Build Array function to the input of the waveform graph. Resize the plot legend to display the labels and edit them. Similar to the example in the preceding section, the **AutoScale** feature of the Y axis should be disabled and the **Loose Fit** option should be unchecked along the X axis.

Place a Wait (ms) function (**Functions » Programming » Timing » Wait**) inside the While Loop to delay the execution in case the VI runs too fast. Right-click on the milliseconds to wait terminal and choose **Create » Constant** from the shortcut menu to create and wire a Numeric Constant. Enter 200 in the box created.

Figure L2-9 and Figure L2-10 illustrate the BD and FP of the designed signal generation system, respectively. Save the VI as *Lab02_ Regular_Waveform.vi* and run it. Change the signal type, gain, and frequency values to see the original and scaled signal in the waveform graph.

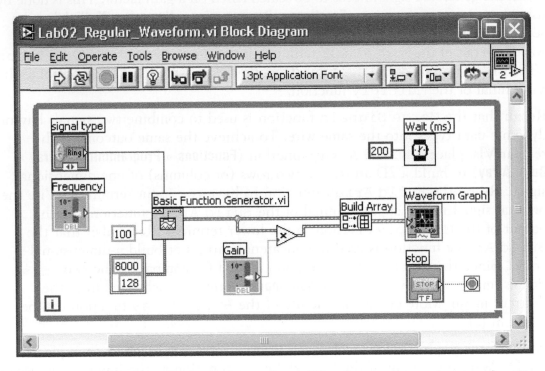

Figure L2-9: BD of signal generation and amplification system using regular VIs.

The waveform data type is not accepted by all the functions or subVIs. To cope with this issue, one extracts the Y component (data value) of the waveform data type to have the output signal as an array of data samples. This is done by placing a Get Waveform Components function (**Functions » Programming » Waveform » Get Waveform Components**). Then, wire the signal out terminal of the Basic Function Generator VI to the waveform terminal of the Get

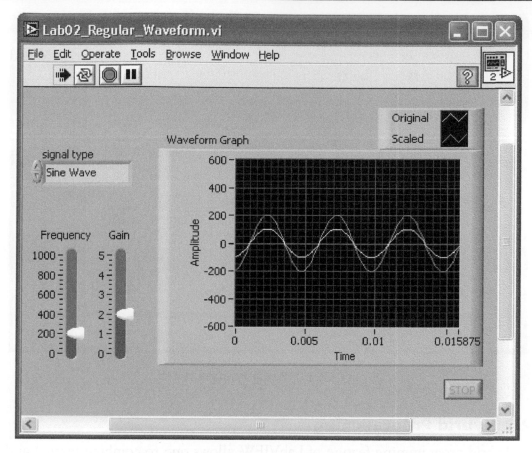

Figure L2-10: Original and scaled output signals.

Waveform Components function. Click on t0, the default terminal, of the Get Waveform Components function and choose Y as the output to extract data values from the waveform data type, as shown in Figure L2-11. The remaining steps are the same as those done for the version shown in Figure L2-9. In this version, however, the processed signal is an array of double precision samples.

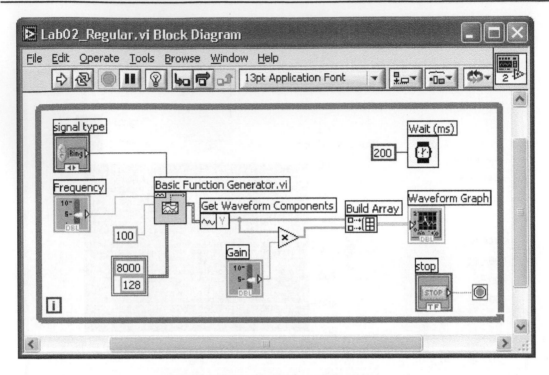

Figure L2-11: Matching data types.

L2.2 Hybrid Programming

The hybrid programming feature of LabVIEW allows one to combine textual and graphical programming, leading to a higher code writing efficiency as compared to pure textual or graphical approaches. In this section, the hybrid programming approach is introduced by redesigning the VIs built earlier using the MathScript and the Call Library Function features of LabVIEW.

L2.2.1 MathScript Feature

In many cases, it is easier to perform math operations via M-files, while carrying out user interfacing, interactivity, and analysis in the more intuitive graphical environment of LabVIEW. M-file codes can be typed in or copied and pasted to MathScript windows or nodes.

Let us now build a program to perform the summation and averaging operations via MathScripting (refer to Figure L2-12). Choose **Functions » Programming » Structures » MathScript** to create a MathSript window. This window appears as a blue box in the BD, and its size can be adjusted if needed. The inputs consist of x and y. To

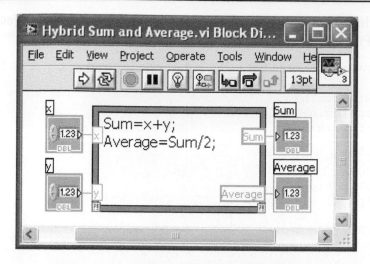

Figure L2-12: BD with MathScript window.

add these inputs, right-click on the border of the MathScript window and click on the Add Input option. After one adds these inputs, two controls are created which allow changing the inputs interactively via the FP. When one right-clicks on the border, the outputs can be added in a similar manner. Also, two numeric indicators are added and wired to the outputs. Although inputs and outputs can be added anywhere on the border, it is a good practice to arrange them such that they match the data flow of a BD. Finally, the M-file code is typed in the MathScript window to describe the relationship between the inputs and the outputs.

L2.2.2 Call Library Function Feature

In this section, the same system is implemented in a hybrid fashion using the Call Library Function Node of LabVIEW. First, a C DLL is built that scales the input signal, and then a VI is built that calls this DLL during run time.

L2.2.2.1 Building C DLL Using MS Visual Studio

After starting the MS Visual Studio, create a new project by selecting **File » New » Project**.... Under **Project types**, choose **Visual C++ » Win32**, and select **Visual Studio installed templates » Win32 Project** under **Templates**. Name the project *Scale* and select a desired directory. Clicking OK opens a Win32 Application Wizard window. Under

Application Settings, choose **DLL** for **Application type**, and check **Empty project** for **Additional options**. Click Finish to complete creating an empty project.

Type the following C source file using any text editor and save it as *Scale.c:*

```c
#include <windows.h>
#include <string.h>
#include <ctype.h>

BOOL WINAPI DllMain (
    HANDLE hModule,
    DWORD dwFunction,
    LPVOID lpNot)
{
    return TRUE;
}

/* This function scales an input signal */
_declspec (dllexport) double Scale(double *signal, double gain)
{
    int i;

    for (i=0; i<28; i++)
        {
            signal[i]=signal[i]*gain;
        }
        return 0;
}
```

Add the C file to the project by selecting **Project » Add Existing Item...**, and build the DLL by selecting **Build » Build Solution.** Now, one can see a file named *Scale.dll* under the *debug* folder of the project directory.

L2.2.2.2 Calling C DLL from LabVIEW

With some small modifications to the VIs built earlier, the DLL built in Visual Studio can get called from LabVIEW. As shown in Figure L2-13, first replace the Multiply function with the Call Library Function Node VI (**Functions » Connectivity » Libraries and Executables » Call Library Function Node**). By default, this VI has no terminals, and it needs to be configured appropriately before it is used.

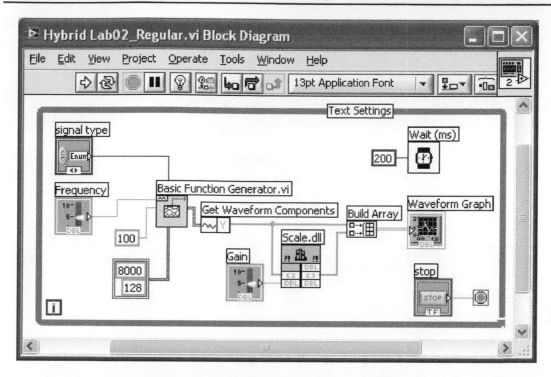

Figure L2-13: Calling DLL from LabVIEW.

Double-clicking this VI brings up a `Call Library Function` window. Under **Function** tab, set **Library name or path** by browsing to the directory of the DLL file (*Scale.dll*) created earlier and set **Function name** to `Scale,` which is the name of the function defined in the source code. Select **Run in UI thread** for **Thread** and **C** for **Calling convention.** Under the tab **Parameters**, the input and output of this VI need to be defined according to the function parameters defined in the C source code. The four buttons ⊞ ⊠ ⬆ ⬇ allow one to add, delete, and reorder parameters. By default, the first parameter is return type, and thus two more input parameters for signal and gain need to be added. The configuration of the parameters is performed as shown in Figure L2-14.

Once this configuration is completed, connect the terminals of the VI as illustrated in Figure L2-13. When this VI is run, the FP should produce the same result as displayed in Figure L2-10.

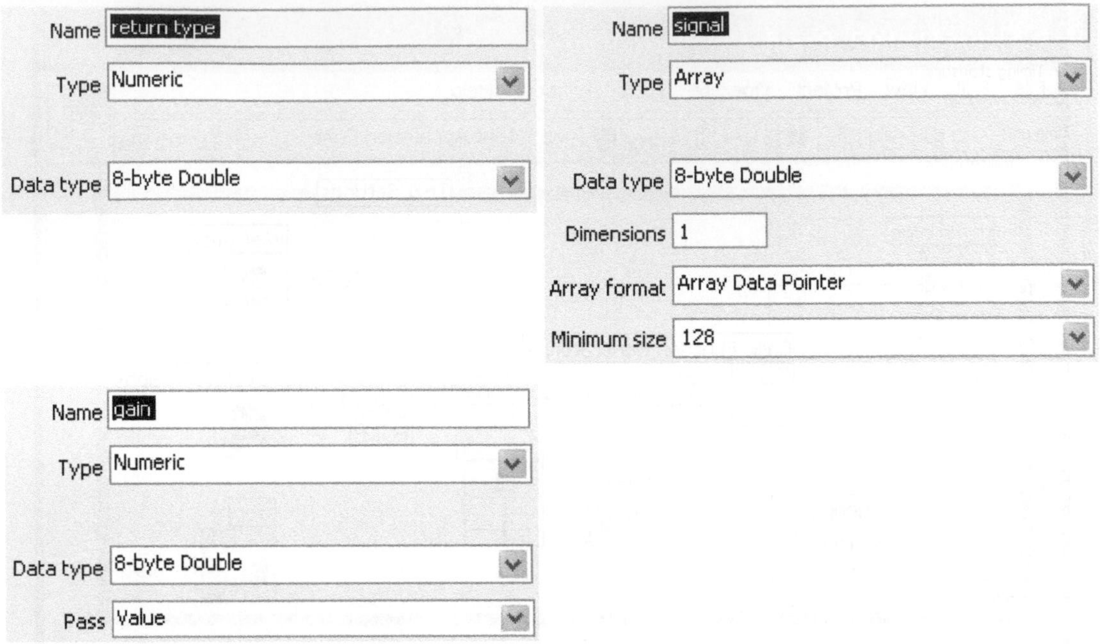

Name | return type |
Type | Numeric |

Data type | 8-byte Double |

Name | signal |
Type | Array |

Data type | 8-byte Double |
Dimensions | 1 |
Array format | Array Data Pointer |
Minimum size | 128 |

Name | gain |
Type | Numeric |

Data type | 8-byte Double |
Pass | Value |

Figure L2-14: Configuration of calling library function parameters.

L2.3 Profile VI

The Profile tool is used to gather timing and memory usage information. Make sure the VI is stopped before setting up a Profile window. Select **Tools » Profile » Performance and Memory** ... to bring up a Profile window.

Place a checkmark in the **Timing Statistics** checkbox to display timing statistics of the VI. The **Timing Details** option provides more detailed statistics of the VI such as drawing time. To profile memory usage as well as timing, check the **Memory Usage** checkbox after checking the **Profile Memory Usage** checkbox. Note that this option can slow down the execution of the VI. Start profiling by clicking the Start button on the profiler; then run the VI. A snapshot of the profiler information can be obtained by clicking on the Snapshot button. After viewing the timing information, click the Stop button. The profile statistics can be stored into a text file by clicking the Save button.

An outcome of the profiler is exhibited in Figure L2-15 after running the Lab02_Regular VI. More details on the use of the Profile tool can be found in [3].

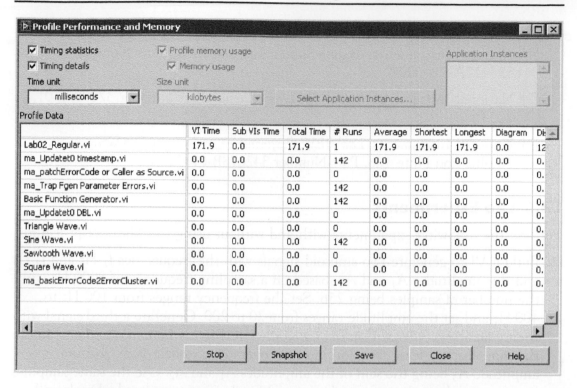

Figure L2-15: Profile window after running Lab02_Regular VI.

L2.4 Bibliography

[1] National Instruments, *Getting Started with LabVIEW*, Part Number 323427A-01, 2003.

[2] National Instruments, *LabVIEW User Manual*, Part Number 320999E-01, 2003.

[3] National Instruments, *LabVIEW Performance and Memory Management*, Application Note 168, Part Number 342078B-01, 2004.

L2.5 Lab Experiments

Perform the following experiments with and without using the MathScript feature.

1. Build a VI to generate two sinusoid signals with the frequencies f_1 Hz and f_2 Hz and the amplitudes A_1 and A_2, based on a sampling frequency of 8000 Hz with the number of samples being 256. Set the frequency ranges from 100 Hz to 400 Hz and set the amplitude ranges from 20 to 200. Generate a third signal with the frequency $f_3 = (mod\ (lcm\ (f_1, f_2),\ 400) + 100)$ Hz, where *mod* and *lcm* denote the modulus and least common multiple operation, respectively, and the amplitude A_3 being the sum of the amplitudes A_1 and A_2. Use the same sampling frequency and number of samples as used for the first two signals. Display all the signals using the legend on the same waveform graph and label them accordingly. When the MathScript feature is not being used, it is easier to use Express VIs.

2. Build a VI to check whether a given positive integer n is a prime number and display a warning message if it is not a prime number.

3. Build a VI to generate two sinusoid signals, the same as the ones in Experiment 2. Generate a third signal with the frequency $f_3 = (gcd\ (f_1, f_2) + mean\ (f_1, f_2))$ Hz, where *gcd* and *mean* denote the greatest common divisor and the average operation, respectively, and the amplitude A_3 being the sum of the amplitudes A_1 and A_2. Use the same sampling frequency and number of samples as used for the first two signals. Display all the signals using the legend on the same waveform graph and label them accordingly. When the MathScript feature is not being used, it is easier to use Express VIs.

4. Build a VI to generate the first n prime numbers and store them using an indexing array. Display the outcome.

Analog-to-Digital Signal Conversion

The process of analog-to-digital signal conversion consists of converting a continuous time and amplitude signal into discrete time and amplitude values. Sampling and quantization constitute the steps needed to achieve analog-to-digital signal conversion. To minimize any loss of information that may occur as a result of this conversion, one must understand the underlying principles behind sampling and quantization.

3.1 Sampling

Sampling is the process of generating discrete time samples from an analog signal. First, it is helpful to mention the relationship between analog and digital frequencies. Let us consider an analog sinusoidal signal $x(t) = A\cos(\omega t + \phi)$. Sampling this signal at $t = nT_s$, with the sampling time interval of T_s, generates the discrete time signal

$$x[n] = A\cos(\omega nT_s + \phi) = A\cos(\theta n + \phi), \qquad n = 0, 1, 2, \ldots, \qquad (3.1)$$

where $\theta = \omega T_s = \dfrac{2\pi f}{f_s}$ denotes digital frequency with units being radians

(as compared to analog frequency ω with units being radians/sec).

The difference between analog and digital frequencies becomes more evident by observing that the same discrete time signal is obtained from different continuous time signals if the product ωT_s remains the same. (An example is shown in Figure 3-1.) Likewise, different discrete time signals are obtained from the same analog or continuous time signal when the sampling frequency is changed. (An example is shown in Figure 3-2.) In other words, both the frequency of an analog signal f and the sampling frequency f_s define the frequency of the corresponding digital signal θ.

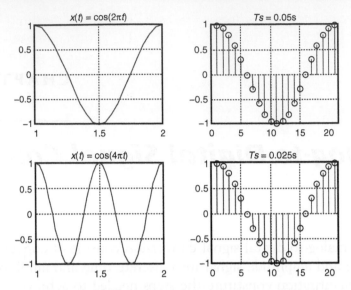

Figure 3-1: Sampling of two different analog signals leading to the same digital signal.

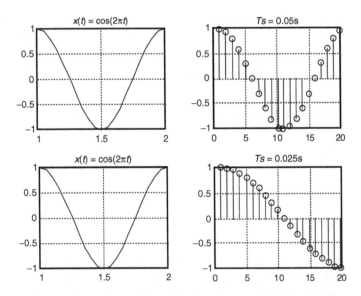

Figure 3-2: Sampling of the same analog signal leading to two different digital signals.

It helps to understand the constraints associated with the preceding sampling process by examining signals in the frequency domain. The Fourier transform pairs in the analog and digital domains are given by

Fourier transform pair for analog signals

$$
\begin{cases}
X(j\omega) = \displaystyle\int_{-\infty}^{\infty} x(t)e^{-j\omega t}dt \\[4mm]
x(t) = \dfrac{1}{2\pi}\displaystyle\int_{-\infty}^{\infty} X(j\omega)e^{j\omega t}d\omega
\end{cases}
\qquad (3.2)
$$

Fourier transform pair for discrete signals

$$
\begin{cases}
X(e^{j\theta}) = \displaystyle\sum_{n=-\infty}^{\infty} x[n]e^{-jn\theta}, \quad \theta = \omega T_s \\[4mm]
x[n] = \dfrac{1}{2\pi}\displaystyle\int_{-\pi}^{\pi} X(e^{j\theta})e^{jn\theta}d\theta
\end{cases}
\qquad (3.3)
$$

As illustrated in Figure 3-3, when an analog signal with a maximum bandwidth of W (or a maximum frequency of f_{max}) is sampled at a rate of $T_s = \dfrac{1}{f_s}$, its

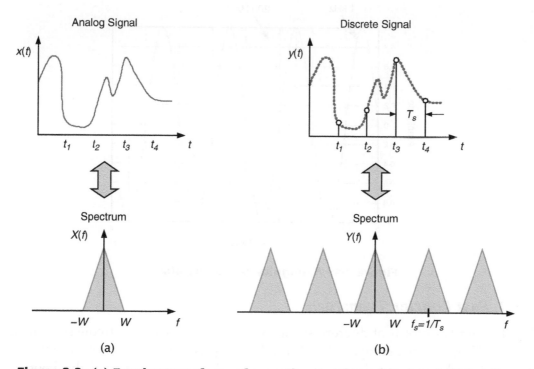

Figure 3-3: (a) Fourier transform of a continuous-time signal and (b) its discrete time version.

corresponding frequency response is repeated every 2π radians, or f_s. In other words, the Fourier transform in the digital domain becomes a periodic version of the Fourier transform in the analog domain. That is why, for discrete signals, one is interested only in the frequency range $[0, f_s/2]$.

Therefore, in order to avoid any aliasing or distortion of the frequency content of the discrete signal, and hence to be able to recover or reconstruct the frequency content of the original analog signal, we must have $f_s \geq 2 f_{max}$. This is known as the Nyquist rate; that is, the sampling frequency should be at least twice the highest frequency in the signal. Normally, before any digital manipulation, a front-end antialiasing lowpass analog filter is used to limit the highest frequency of the analog signal.

The aliasing problem can be further illustrated by considering an undersampled sinusoid, as depicted in Figure 3-4. In this figure, a 1 kHz sinusoid is sampled at $f_s = 0.8$ kHz, which is less than the Nyquist rate of 2 kHz. The dashed-line signal is a 200 Hz sinusoid passing through the same sample points. Thus, at the sampling frequency of 0.8 kHz, the output of an A/D converter would be the same if either of the 1 kHz or 200 Hz sinusoids was the input signal. On the other hand, oversampling a signal provides a richer description than that of the signal sampled at the Nyquist rate.

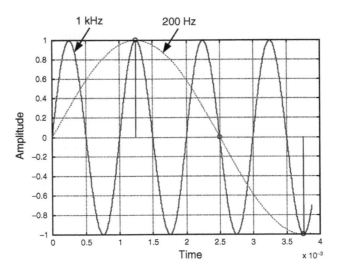

Figure 3-4: Ambiguity caused by aliasing.

3.1.1 Fast Fourier Transform

The Fourier transform of discrete signals is continuous over the frequency range $[0, f_s/2]$. Thus, from a computational standpoint, this transform is not suitable to use. In practice, the discrete Fourier transform (DFT) is used in place of the Fourier transform. DFT is analogous with Fourier series in the analog domain. Detailed descriptions of signal transforms can be found in various textbooks on digital signal processing, e.g., [1], [2]. Fourier series and DFT transform pairs are expressed as

Fourier series for periodic analog signals

$$\begin{cases} X_k = \dfrac{1}{T}\displaystyle\int_{-T/2}^{T/2} x(t)e^{-j\omega_0 kt}dt \\[4mm] x(t) = \displaystyle\sum_{k=-\infty}^{\infty} X_k e^{j\omega_0 kt} \end{cases} \qquad (3.4)$$

where T denotes period and ω_0 fundamental frequency.

Discrete Fourier transform (DFT) for periodic discrete signals

$$\begin{cases} X[k] = \displaystyle\sum_{n=0}^{N-1} x[n]e^{-j\frac{2\pi}{N}nk}, \qquad k = 0, 1, \ldots, N-1 \\[4mm] x[n] = \dfrac{1}{N}\displaystyle\sum_{k=0}^{N-1} X[k]e^{j\frac{2\pi}{N}nk}, \qquad n = 0, 1, \ldots, N-1 \end{cases} \qquad (3.5)$$

It should be noted that DFT and Fourier series pairs are defined for periodic signals. Hence, when one is computing DFT, it is required to assume periodicity with a period of N samples. Figure 3-5 illustrates a sampled sinusoid which is no longer

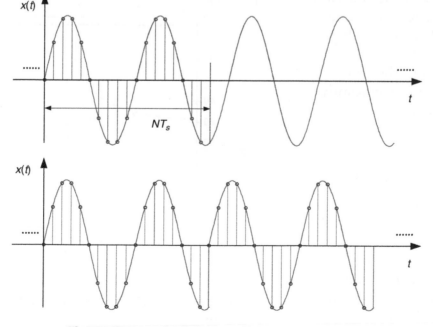

Figure 3-5: Periodicity condition of sampling.

periodic. In order to make sure that the sampled version remains periodic, the analog frequency should satisfy this condition [3]

$$f = \frac{m}{N} f_s \qquad (3.6)$$

where m denotes the number of cycles over which DFT is computed.

The computational complexity (number of additions and multiplications) of DFT is reduced from N^2 to $N \log N$ by using fast Fourier transform (FFT) algorithms. In these algorithms, N is normally considered to be a power of two. Figure 3-6 shows the effect of the periodicity constraint on the FFT computation. In this figure, the FFTs of two sinusoids with frequencies of 250 Hz and 251 Hz are shown. The amplitudes of the sinusoids are considered to be one. Although there is only a 1Hz difference between the sinusoids, the FFT outcomes are significantly different due to the improper sampling.

3.2 Quantization

An A/D converter has a finite number of bits (or resolution). As a result, continuous amplitude values get represented or approximated by discrete amplitude levels.

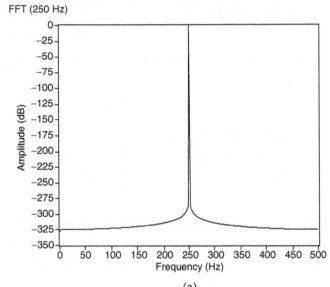

(a)

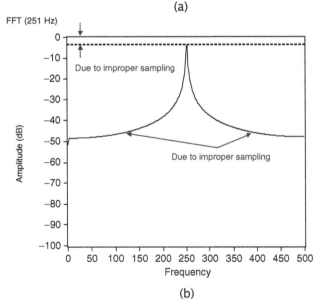

(b)

Figure 3-6: FFTs of (a) a 250 Hz and (b) a 251 Hz sinusoid.

The process of converting continuous into discrete amplitude levels is called quantization. This approximation leads to errors called quantization noise. The input/output characteristic of a 3-bit A/D converter is shown in Figure 3-7 to see how analog voltage values are approximated by discrete voltage levels.

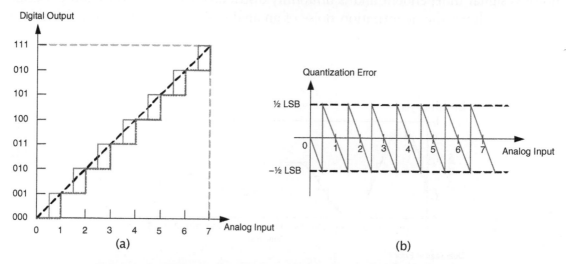

(a)   (b)

Figure 3-7: Characteristic of a 3-bit A/D converter: (a) input/ output transfer function and (b) additive quantization noise.

A quantization interval depends on the number of quantization or resolution levels, as illustrated in Figure 3-8. Clearly, the amount of quantization noise generated by an A/D converter depends on the size of the quantization interval.

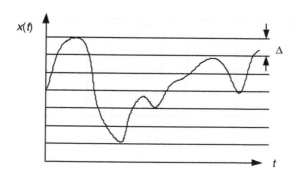

Figure 3-8: Quantization levels.

More quantization bits translate into a narrower quantization interval and hence into a lower amount of quantization noise.

In Figure 3-8, the spacing Δ between two consecutive quantization levels corresponds to one least significant bit (LSB). Usually, it is assumed that quantization noise is signal independent and is uniformly distributed over –0.5 LSB and 0.5 LSB. Figure 3-9 shows the quantization noise of an analog signal quantized by a 3-bit A/D converter.

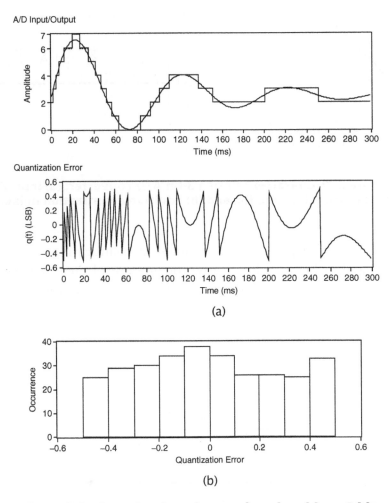

(a)

(b)

Figure 3-9: Quantization of an analog signal by a 3-bit A/D converter: (a) output signal and quantization error, (b) histogram of quantization error, and

Continued

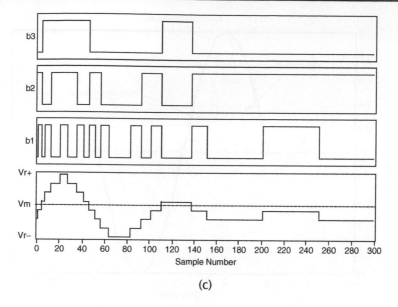

(c)

Figure 3-9 Continued: (c) bit stream.

3.3 Signal Reconstruction

So far, we have examined the forward process of sampling. It is also important to understand the inverse process of signal reconstruction from samples. According to the Nyquist theorem, an analog signal v_a can be reconstructed from its samples by using the following equation:

$$v_a(t) = \sum_{k=-\infty}^{\infty} v_a[kT_s] \left[\mathrm{sinc}\left(\frac{t-kT_s}{T_s}\right) \right] \tag{3.7}$$

One can see that the reconstruction is based on the summations of shifted sinc functions. Figure 3-10 illustrates the reconstruction of a sinewave from its samples.

It is very difficult to generate sinc functions by electronic circuitry. That is why, in practice, a pulse approximation of a sinc function is used. Figure 3-11 shows a sinc function approximated by a pulse, which is easy to realize in electronic circuitry. In fact, the well-known sample and hold circuit performs this approximation [3].

65

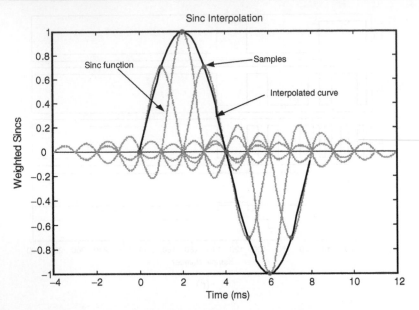

Figure 3-10: Reconstruction of an analog sinewave based on its samples, $f = 125$ Hz, and $f_s = 1$ kHz.

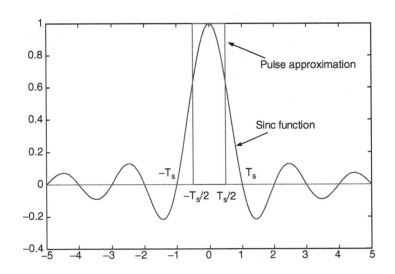

Figure 3-11: Approximation of a sinc function by a pulse.

3.4 Bibliography

[1] J. Proakis and D. Manolakis, *Digital Signal Processing: Principles, Algorithms, and Applications*, Prentice-Hall, 1996.

[2] S. Mitra, *Digital Signal Processing: A Computer-Based Approach*, McGraw-Hill, 2001.

[3] B. Razavi, *Principles of Data Conversion System Design*, IEEE Press, 1995.

Lab 3: Sampling, Quantization, and Reconstruction

This lab covers several examples to further convey sampling, quantization, and reconstruction aspects of analog-to-digital and digital-to-analog signal conversion processes.

L3.1 Aliasing

In this example, a discrete signal is generated by sampling a sinusoidal signal. When the normalized frequency f/f_s of the discrete signal becomes greater than 0.5, or the Nyquist frequency, the aliasing effect becomes evident.

A sampling process is done by setting the sampling frequency f_s to 1 kHz, and the number of samples N to 10. This results in a 10 ms sampled signal. The signal frequency is arranged to vary between 0 and 1000 Hz using an FP control. Figure L3-1 shows a sinusoidal signal having a frequency of 300 Hz which is sampled at 1 kHz for 10 ms producing 10 samples, which are displayed in a waveform graph. In this graph, an analog signal representation is also made by oversampling the sinusoidal signal 100 times faster. In other words, an analog signal representation is obtained by considering a sampling frequency of 100 kHz generating 1000 samples.

The FP of the VI includes a Horizontal Pointer Slide control for the signal frequency and two Numeric Indicators for the normalized frequency and aliased frequency. A Stop Button associated with a While Loop on the BD is located on the FP. This button is used to stop the execution of the VI.

Figure L3-2 shows the BD for this sampling system. To generate the analog and discrete sinusoids, one uses a MathScript node. The input to the script node consists of frequency f and sampling frequency f_s. The output of the script node consists of three arrays corresponding to the sinusoidal samples (discrete, analog, and aliased) and normalized aliased frequency f_a. The Build Waveform function (**Functions » Programming » Waveform » Build Waveform**) is used to build a waveform by combining the samples into the Y terminal, and the time duration between samples, $T_s=1/1000$, into the dt terminal. As discussed earlier, the number of samples for the analog representation of the signal is considered to be 100 times that of the discrete signal. Thus, the time interval of the analog signal is set to one hundredth of that of the discrete signal.

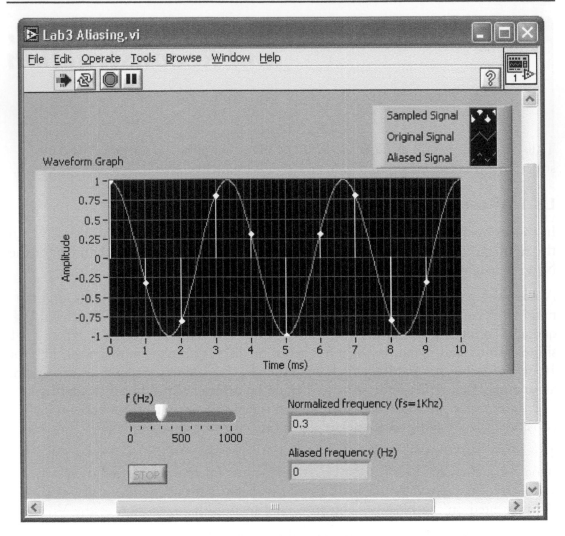

Figure L3-1: Aliasing effect.

In the MathScript node, the discrete signal and the analog signal are generated by sampling a cosine wave at 1 kHz and 100 kHz, respectively, and the aliased signal is generated when the signal frequency gets higher than the Nyquist frequency.

An `If-else Statement` is used to handle the sampling cases with and without aliasing. If the normalized frequency `fn` is greater than 0.5, the aliased frequency is calculated and used for generating the aliased signal. If `fn` is less than or equal to 0.5, the aliased signal is the same as the analog signal, and the aliased frequency is set to 0.

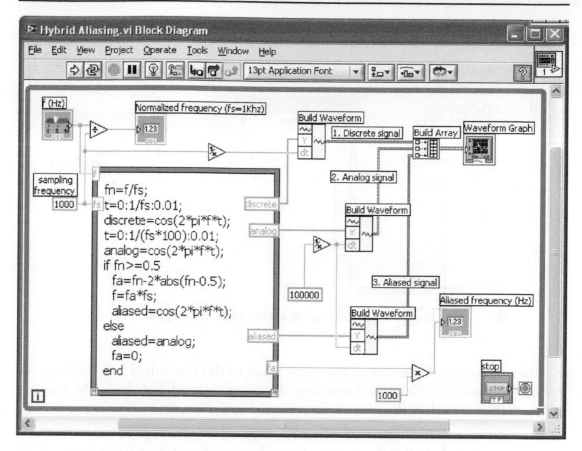

Figure L3-2: BD of aliasing example using MathScript.

When using a MathScript node, one needs to set the input/output data type manually corresponding to the variables in the script. The default data type is set to Scalar Double, which needs to be changed to 1-D Array Double to get the output as a signal. This can be done by right-clicking on any of the output nodes and selecting **Choose Data Type** (see Figure L3-3).

Alternatively, instead of using MathScript, one can use Sine Wave VIs (**Functions » Signal Processing » Signal Generation » Sine Wave**) to generate the analog and discrete signals. These VIs are arranged vertically in the middle of the BD shown in Figure L3-4. The inputs to these VIs comprise number of samples, amplitude, frequency, and phase offset. Amplitude is set to 1 by default in the absence of any wiring to the amplitude terminal. The f terminal requires frequency to be specified in cycles per sample, which is the reciprocal of number of samples per period.

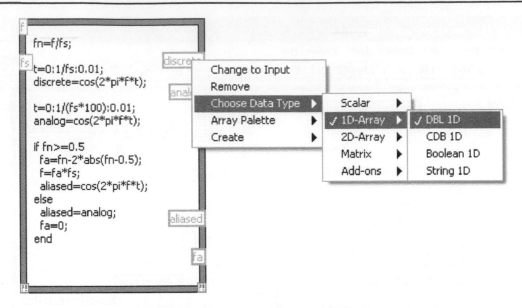

Figure L3-3: Changing script node data type.

For the analog signal generation, the value wired to the f terminal is divided by 100 because it is sampled 100 times faster than the discrete signal. For phase, the numeric constant 90 is wired to the phase in terminal.

Among the three Sine Wave VIs shown in Figure L3-4, the top VI generates the discrete signal, the middle VI generates the analog signal, and the bottom VI generates the aliased signal when the signal frequency gets higher than the Nyquist frequency.

A Case Structure is used to handle the sampling cases with aliasing and without aliasing. If the normalized frequency is greater than 0.5, corresponding to the True case, the third Sine Wave VI generates an aliased signal. All the inputs except for the aliased signal frequency are the same.

Note that an Expression Node (**Functions » Programming » Numeric » Expression Node**) is used to obtain the aliased frequency. An Expression Node is usually used to calculate an expression of a single variable. Many built-in functions, e.g., abs (absolute), can be used in an Expression Node to evaluate an equation. More details on the use of Expression Node can be found in [1].

For the False case, i.e., sampling without aliasing, there is no need to generate an aliased signal. Thus, the analog signal is connected to the output of the Case Structure so that the same signal is drawn on the waveform graph and

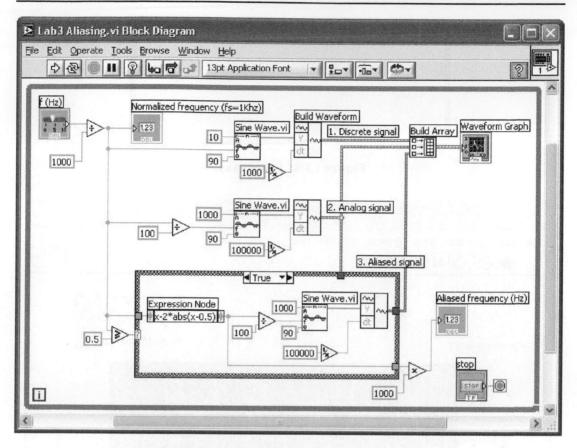

Figure L3-4: BD of aliasing example—True case.

the frequency of the aliased signal is set to 0. This is illustrated in Figure L3-5. It should be remembered that, when using a `Case Structure`, one needs to wire all the outputs for each case.

An aliasing outcome is illustrated in Figure L3-6, where samples of a 700 Hz sinusoid are shown. Note that these samples could have also been obtained from a 300 Hz sinusoid, shown by the dotted line in Figure L3-6.

Next, all the three waveforms are bundled together by using the `Build Array` function and displayed in the same graph. The properties of the waveform graph should be configured as shown in Figure L3-7. This is done by expanding the plot legend vertically to display the three entries and renaming the labels appropriately. Right-click on the waveform graph and choose **Properties** from the shortcut menu.

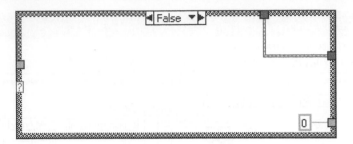

Figure L3-5: False case.

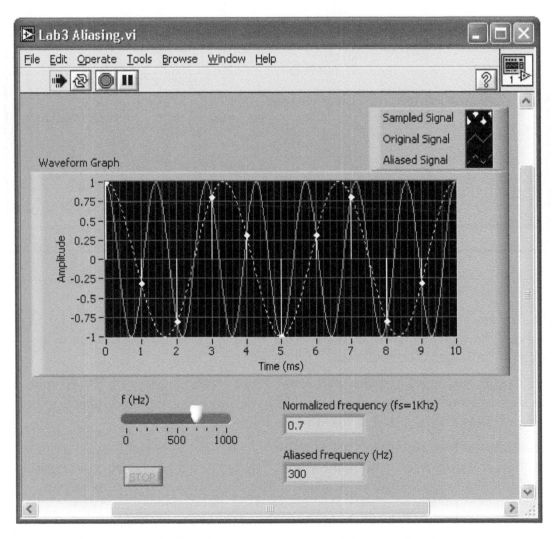

Figure L3-6: A 700 Hz sinusoid aliased with a 300 Hz sinusoid.

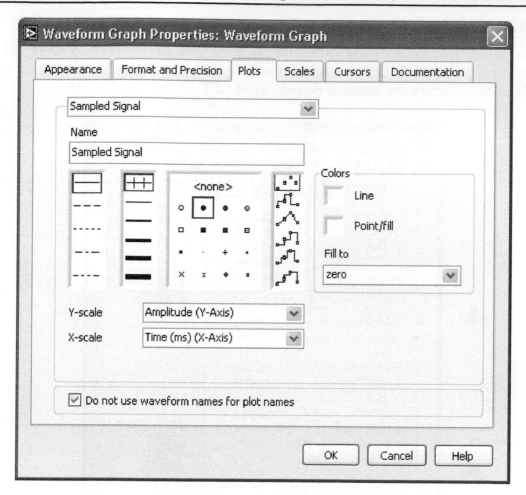

Figure L3-7: Waveform Graph Properties dialog box.

A Waveform Graph Properties dialog box will be brought up. Select the **Plots** tab to modify the plot style. Choose `Sampled Signal` in the **Plot** drop-down menu, as shown in Figure L3-7. Also, choose the options for **Point Style, Plot Interpolation**, and **Fill to** as indicated in this figure. Adjust the line style of the aliased signal to dotted line.

Rename all the controls and indicators, and modify the maximum scale of the `Horizontal Pointer Slide` control to 1000 to complete the VI.

L3.2 Fast Fourier Transform

The analog frequency should satisfy the condition in Equation (3.6) to avoid any discontinuity in DFT. Let us build the example shown in Figure L3-8 using Express VIs to demonstrate the required periodicity of DFT.

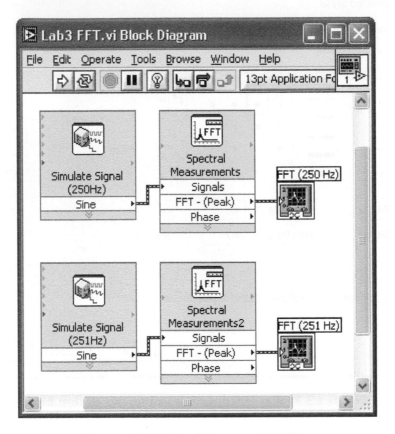

Figure L3-8: BD of Express VI FFTs.

Use two `Simulate Signal` Express VIs (**Functions » Express » Input » Simulate Signal**) to simulate the signals. Placing a `Simulate Signal` Express VI brings up a configuration dialog box for setting up the parameters, including signal type, frequency, amplitude, and sampling frequency, as shown in Figure L3-9. Choose **Sine** for the signal type; then set the frequency to 250, the amplitude to 1, and the phase to 90. Furthermore, enter 1000 as the sampling frequency and 512 as the number of samples. These parameters satisfy the condition in Equation (3.6). As for the 251 Hz sinusoid, use the same parameters except for the frequency, which is to be set to 251.

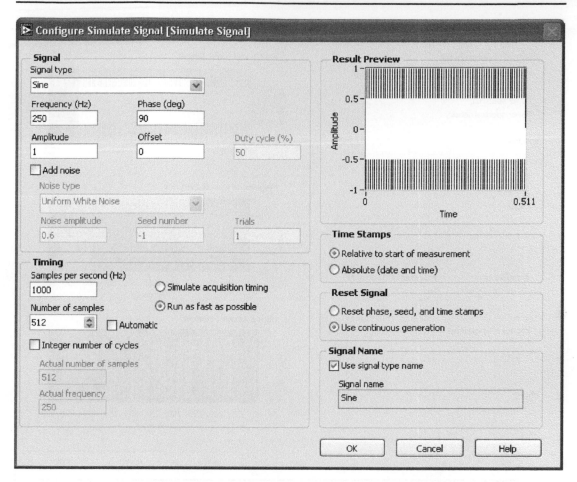

Figure L3-9: Configuration dialog box of Simulate Signal Express VI.

Now, place two Spectral Measurements Express VIs (**Functions »
Express » Signal Analysis » Spectral Measurements**) to compute the FFTs of the
signals. The configuration dialog box entries need to be adjusted as shown in
Figure L3-10. The adjustments shown in this figure provide the spectrum in
dB scale without using a spectral leakage window. Notice that when the
parameters are adjusted, the preview windows are updated based on the current
setting.

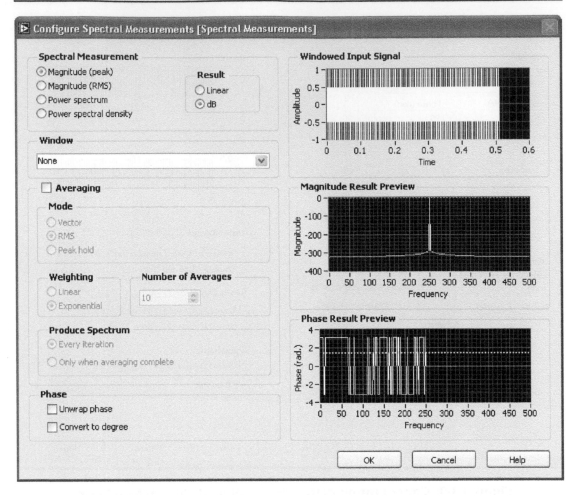

Figure L3-10: Configuration dialog box of Spectral Measurements Express VI.

The spectra of the two signals are shown in Figure L3-11. As seen from this figure, the spectrum of the 251 Hz signal is spread over a wide range due to the improper sampling. Also, its peak drops by nearly 4 dB.

The plot in the waveform graph can be magnified using the **Graph Palette** for better visualization. The **Graph Palette** can be displayed by right-clicking on the waveform graph and choosing **Visible Item » Graph Palette** from the shortcut menu. The options **Cursor Movement Tool, Zoom**, and **Panning Tool** are provided in the palette. More specific options for zooming in and out are available in the expanded menu when the **Zoom** option is chosen as shown in Figure L3-12.

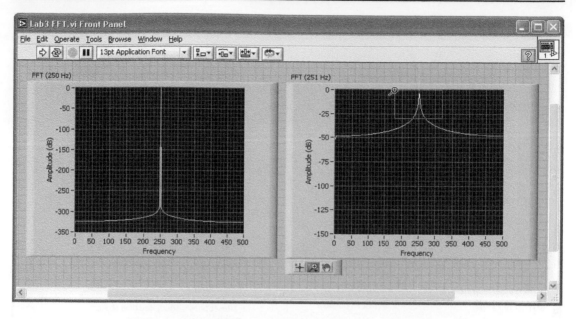

Figure L3-11: FFTs of 250 and 251 Hz sinusoids.

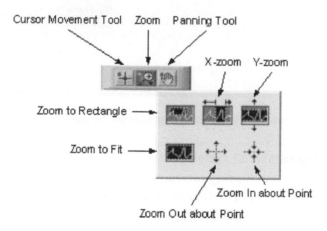

Figure L3-12: Menu options of Graph palette.

The improper sampling for the 251 Hz signal can be corrected by modifying the sampling parameters. The configuration dialog box of the Simulate Signal Express VI provides a useful option, which is **Integer number of cycles**, to satisfy the sampling condition. This is illustrated in Figure L3-13.

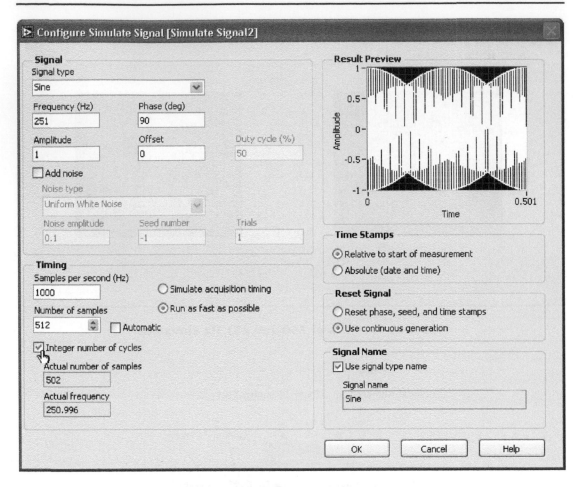

Figure L3-13: Modifying sampling parameters.

Checking the **Integer number of cycles** option alters the number of samples and frequency to 502 and 250.996, respectively. As a result, a proper sampling condition is established. The spectrum of this resampled signal is shown in Figure L3-14. As seen from this figure, the frequency leakage is considerably reduced in this case.

L3.3 Quantization

Let us now build an A/D converter VI to illustrate the quantization effect. An analog signal given by

$$y(t) = 5.2\exp(-10t)\sin(20\pi t) + 2.5$$

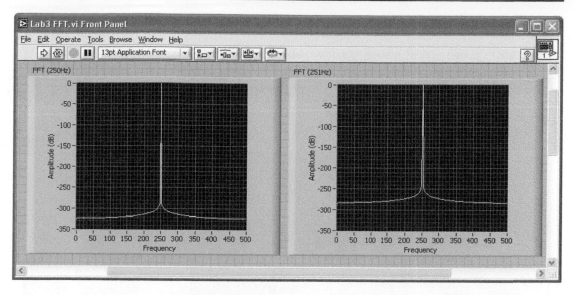

Figure L3-14: FFTs of 250 and 251Hz sinusoids (modified sampling condition).

is considered for this purpose. Note that the maximum and minimum values of the signal fall in the range 0 to 7, which can be represented by 3 bits. On the FP, the quantization error, the histogram of the quantization error, as well as the quantized output are displayed as indicated in Figure L3-15.

To build the converter BD, as shown in Figure L3-16, one needs to use the Formula Waveform VI (**Functions » Programming » Waveform » Analog Waveform » Waveform Generation » Formula Waveform**). The inputs to this VI comprise a string constant specifying the formula, amplitude, frequency, and sampling information. The values of the output waveform, Y component, are extracted with the Get Waveform Components function (**Functions » Programming » Waveform » Get Waveform Components**).

To exhibit the quantization process, one can use the To Unsigned Byte Integer function (**Functions » Programming » Numeric » Conversion » To Unsigned Byte Integer**) to convert the double precision signal into an unsigned integer signal. The resolution of quantization is assumed to be 3 bits, noting that the amplitude of the signal remains between 0 and 7. Values of the analog waveform are replaced by quantized values forming a discretized waveform. This is done by wiring the quantized values to a Build Waveform function while the other properties are kept the same as the analog waveform.

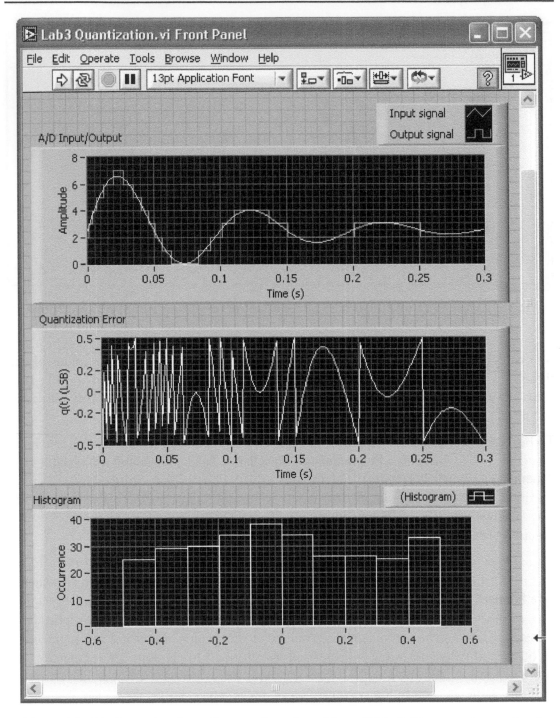

Figure L3-15: Quantization of an analog signal by a 3-bit A/D converter: output signal, quantization error, and histogram of quantization error.

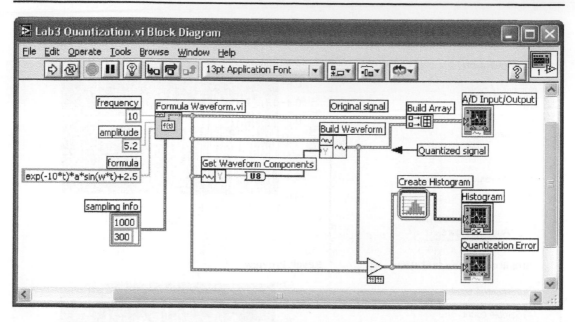

Figure L3-16: Quantization of an analog signal by a 3-bit A/D converter.

Now the difference between the input and quantized output values can be found by using the Subtract function. This difference represents the quantization error. Also, the histogram of the quantization error is obtained by using the Create Histogram Express VI (**Functions » Express » Signal Analysis » Create Histogram**). Placing this VI brings up a configuration dialog, as shown in Figure L3-17. The maximum and minimum quantization errors are 0.5 and –0.5, respectively. Hence, the number of bins is set to 10 in order to divide the errors between –0.5 and 0.5 into 10 levels. In addition, for the **Amplitude Representation** option, choose **Sample count** to generate the histogram. A waveform graph can be created by right-clicking on the Histogram node of the Create Histogram Express VI and choosing **Create » Graph Indicator**.

Return to the FP and change the property of the graph for a more understandable display of the discrete signal. Add the plot legend to the waveform graph and resize it to display the two signals. Rename the analog signal as Input Signal and the discrete signal as Output Signal.

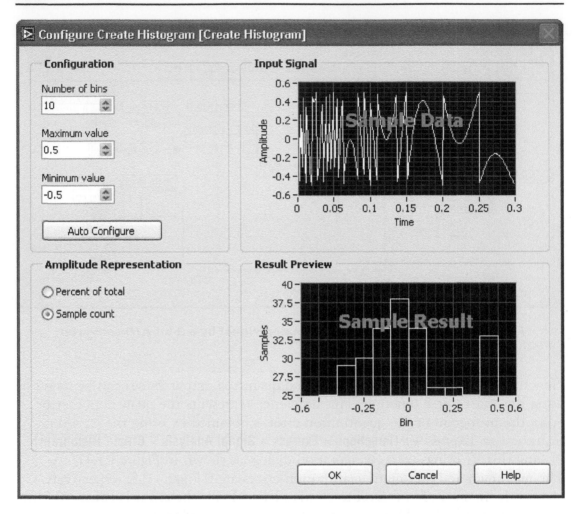

Figure L3-17: Configuration dialog box of Create Histogram Express VI.

To display the discrete signal, bring up the Properties dialog box by right-clicking and choosing **Properties** from the shortcut menu. Click the **Plots** tab and choose the signal plot `Output Signal`. Then, choose **stepwise horizontal**, indicated by ⌐⌐, from the **Plot Interpolation** option as the interpolation method. Now, the VI is complete, as shown in Figure L3-18.

Next, let us build a VI which can analyze the quantized discrete waveform into a bitstream resembling a logic analyzer. For a 3-bit A/D converter, the bitstream can be represented by $b_3 b_2 b_1$ in binary format. The discrete waveform and its bit decomposition are shown in Figure L3-18.

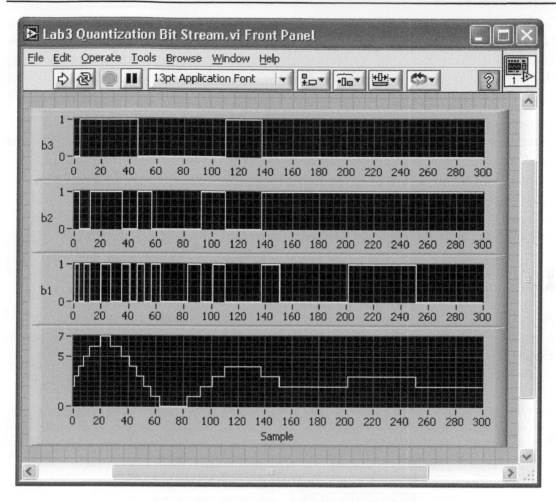

Figure L3-18: Bitstream of 3-bit quantization.

The same analog signal used in the previous example is considered here. The analog signal is generated by a Formula Waveform VI and quantized by using a To Unsigned Byte Integer function. Locate a For Loop to repeat the quantization as many times as the number of samples. This number is obtained by using the Array Size function (**Functions » Programming » Array » Array Size**). Wire this number to the Count terminal of the For Loop.

Wiring the input array to the For Loop places a Loop Tunnel on the loop border. Note that auto-indexing is enabled by default when one inputs an array into a For Loop. With auto-indexing enabled, each element of the input array is passed into the loop one at a time per loop iteration.

In order to obtain a binary bitstream, each value passed into the For Loop is converted into a Boolean array via a Number To Boolean Array function (**Functions » Programming » Boolean » Number To Boolean Array**). The elements of the Boolean array represent the decomposed bits of the 8-bit integer. The value of a specific bit can be accessed by passing the Boolean array into an Index Array function (**Functions » Programming » Array » Index Array**) and specifying the bit location with a Numeric Constant. Since the values stored in the array are Boolean, i.e. false or true, they are then converted into 0 and 1, respectively, using the Boolean To (0,1) function (**Functions » Programming » Boolean » Boolean To (0,1)**). Data from each bit location are wired out of the For Loop. Note that an array output is created with the auto-indexing being enabled.

As configured in the previous example, the **stepwise horizontal** interpolation method is used for the waveform graph of the discrete signal. The completed VI is shown in Figure L3-19.

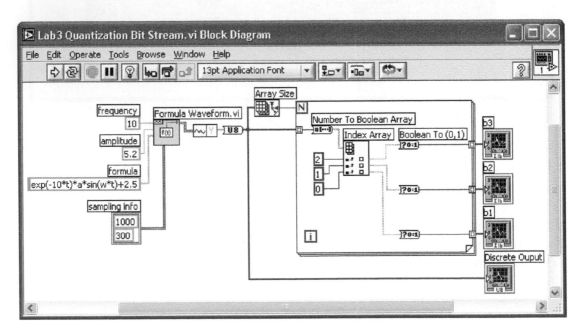

Figure L3-19: Logic analyzer BD.

86

L3.4 Signal Reconstruction

As the final example in this lab, a signal reconstruction VI is built. Let us examine the FP shown in Figure L3-20 exhibiting a sampled signal and its reconstructed version. The reconstruction kernel is also shown in this FP.

The sampled signal is shown via bars in the top waveform graph. In order to reconstruct an analog signal from the sampled signal, a convolution operation with a sinc function is carried out as specified by Equation (3.7).

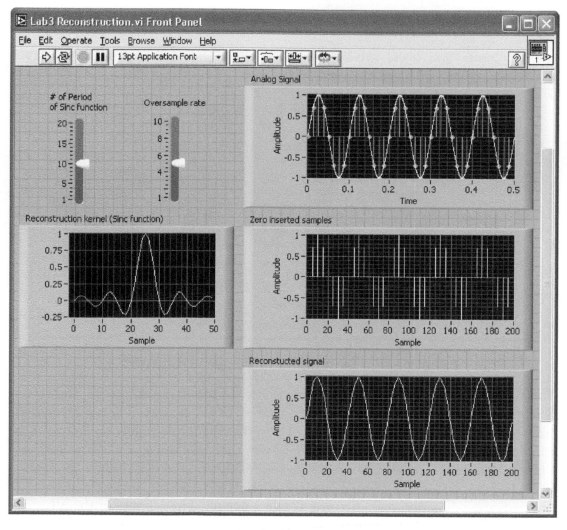

Figure L3-20: FP of a reconstructed sinewave from its samples.

Let us now build the VI. It is assumed that a unity amplitude sinusoid of 10 Hz is sampled at 80 Hz. To display the reconstructed analog signal, one sets the sampling frequency and number of samples to 100 times those of the discrete signal. The two waveforms are merged and displayed in the same waveform graph as shown in Figure L3-20.

The BD of the signal reconstruction system is shown in Figure L3-21. Two custom subVIs are shown on this BD. The Add Zeros VI is used to insert zeros between consecutive samples to simulate oversampling, and the Sinc Function VI is used to generate samples of a sinc function with a specified number of zero-crossings.

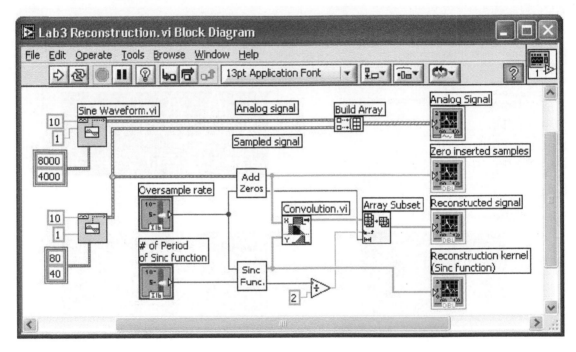

Figure L3-21: BD of signal reconstruction system.

The BD of each subVI is briefly explained here. In the Add Zeros VI, shown in Figure L3-22, zero rows are concatenated to the 1D signal array. The augmented 2D array is then transposed and reshaped to 1D so that the zeros are located between the samples. The number of zeros inserted between the samples can be controlled by wiring a numeric control. The output waveform shown in the BD takes its input from the other VI and is created by right-clicking on the Get Waveform Components function and choosing **Create » Control**. The outputs of the VI comprise the array of zero-inserted samples and the total number of samples. The connector pane of the VI consists of two input terminals and two output terminals.

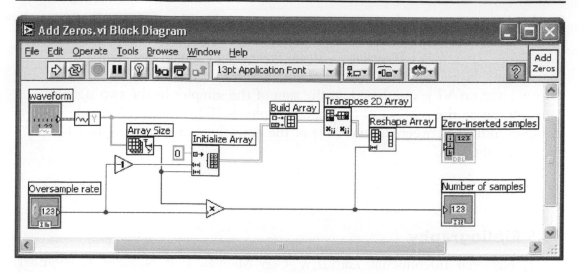

Figure L3-22: Add Zeros subVI.

The input terminals are wired to the controls and the output terminals to the indicators, respectively.

The Sinc Function VI, shown in Figure L3-23, generates samples of a sinc function based on the number of samples, delay, and sampling interval parameters.

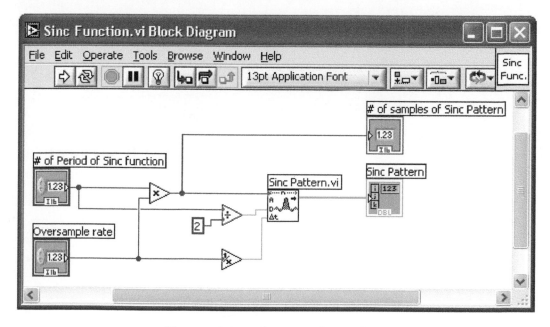

Figure L3-23: Sinc Function subVI.

Finally, let us return to the BD shown in Figure L3-21. The two signals generated by the subVIs, i.e., the zero-inserted signal and sinc signal, are convolved using the `Convolution` VI (**Functions » Signal Processing » Signal Operation » Convolution**). Note that the length of the convolved array obtained from the `Convolution` VI is one less than the sum of the samples in the two signals, e.g., 249. Since the number of the input samples is 200, only a 200 sample portion (samples indices between 25 and 224) of the convolved output is displayed for better visualization.

L3.5 Bibliography

[1] National Instruments, *LabVIEW User Manual*, Part Number 320999E-01, 2003.

L3.6 Lab Experiments

Perform the following experiments with and without using the MathScript feature.

1. Build a VI to generate the signal given by Equation (3.8) with the frequency f Hz and amplitude A based on a sampling frequency of 4000 Hz with the number of samples being 200. Set the frequency range from 1 Hz to 1000 Hz and the amplitude range from 0 to 25. Generate the quantized bit stream and display it together with the quantization error. Also, reconstruct an analog signal using the above-sampled signal by performing a convolution operation with a sinc function as specified by Equation (3.7).

$$x(t) = A^{(3/2)} \sin (2\pi f t) + \sqrt{3.7} \qquad (3.8)$$

2. Build a VI to compute m, the number of cycles over which DFT must be computed, as indicated by Equation (3.9) with analog frequency f, sampling frequency f_s, and total number of samples N. Specify the data types of the controls—analog frequency, sampling frequency, and total number of samples— to be DBL, I32, and I32, respectively. The VI should also issue a warning message in case m is not an integer due to improper sampling.

$$f = \left(\frac{m}{N}\right) f_s \qquad (3.9)$$

3. Build a VI to generate the signal given by Equation (3.10) with the frequencies f_1 Hz and f_2 Hz and the amplitude A with the number of samples being 300. Compute the sampling frequency as $(4 * max (f_1, f_2))$. Set the frequency ranges from 1 Hz to 1 KHz and the amplitude range from 0 to 40. Generate the quantized bit stream and display it together with the quantization error. Also, reconstruct an analog signal using the above-sampled signal by performing a convolution operation with a sinc function as specified by Equation (3.7).

$$x(t) = A \sin (2\pi f_1 t) + A \cos (2\pi f_2 t) \qquad (3.10)$$

4. Build a VI to compute m similar to (2). In case of improper sampling, the VI should compute and display the nearest possible analog frequency f_{new} and total number of samples N_{new} for having an integer number of cycles m_{new}. In the absence of improper sampling, use $f_{new} = f$, $N_{new} = N$, and $m_{new} = m$. **Hint:** If m is not an integer, then round it to the nearest integer m_{new} and then recompute N_{new}. If N_{new} is not an integer, round it to the nearest integer and compute f_{new} by using the updated values of m_{new} and N_{new} for the specified sampling frequency.

Digital Filtering

Filtering of digital signals is a fundamental concept in digital signal processing. Here, it is assumed that the reader has already taken a theory course in digital signal processing or is already familiar with Finite Impulse Response (FIR) and Infinite Impulse Response (IIR) filter design methods.

In this chapter, the structure of digital filters is briefly mentioned, followed by a discussion on the LabVIEW Digital Filter Design (DFD) toolkit. This toolkit provides various tools for the design, analysis, and simulation of digital filters.

4.1 Digital Filtering

4.1.1 Difference Equations

As a difference equation, an FIR filter is expressed as

$$y[n] = \sum_{k=0}^{N} b_k x[n-k] \qquad (4.1)$$

where b's denote the filter coefficients and N the number of zeros or filter order. As described by this equation, an FIR filter operates on a current input $x[n]$ and a number of previous inputs $x[n-k]$ to generate a current output $y[n]$.

The equi-ripple method, also known as the Remez algorithm, is normally used to produce an optimal FIR filter [1]. Figure 4-1 shows the filter responses using the available design methods consisting of equi-ripple, Kaiser window, and Dolph-Chebyshev window. Among these methods, the equi-ripple method generates a response whose deviation from the desired response is evenly distributed across the passband and stopband [2].

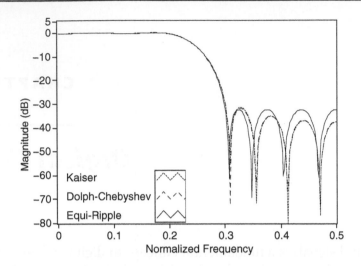

Figure 4-1: Responses of different FIR filter design methods.

The difference equation of an IIR filter is given by

$$y[n] = \sum_{k=0}^{N} b_k x[n-k] - \sum_{k=1}^{M} a_k y[n-k] \qquad (4.2)$$

where b's and a's denote the filter coefficients and N and M the number of zeros and poles, respectively. As indicated by Equation (4.2), an IIR filter uses a number of previous outputs $y[n-k]$ as well as a current and a number of previous inputs to generate a current output $y[n]$.

Several methods are widely used to design IIR filters. They include Butterworth, Chebyshev, Inverse Chebyshev, and Elliptic methods. Figure 4-2 shows the magnitude response of an IIR filter designed by these methods having the same order for comparison purposes. For example, the elliptic method generates a relatively narrower transition band and more ripples in passband and stopband, whereas the Butterworth method generates a monotonic type response [2]. Table 4-1 summarizes the characteristics of these design methods.

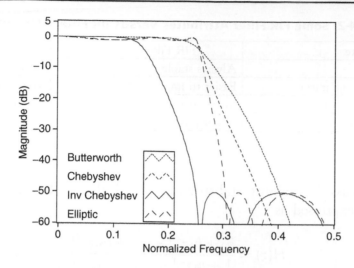

Figure 4-2: Responses of different IIR filter design methods.

Table 4-1: Comparison of Different IIR Filter Design Methods [1]

IIR Filter	Ripple in Passband?	Ripple in Stopband?	Transition Bandwidth	Needed Order for Given Filter Specifications
Butterworth	No	No	Widest	Highest
Chebyshev	Yes	No	Narrower	Lower
Inverse Chebyshev	No	Yes	Narrower	Lower
Elliptic	Yes	Yes	Narrowest	Lowest

4.1.2 Stability and Structure

In general, as compared to IIR filters, FIR filters require less precision and are computationally more stable. The stability of an IIR filter depends on whether its poles are located inside the unit circle in the complex plane. Consequently, when an IIR filter is implemented on a fixed-point processor, its stability can be affected. Table 4-2 provides a summary of the differences between the attributes of FIR and IIR filters.

Table 4-2: Some FIR Filter Attributes Versus IIR Filter Attributes [1]

Attribute	FIR Filter	IIR Filter
Stability	Always stable	Conditionally stable
Fixed-point implementation	Easier to implement	More involved
Computational complexity	More operations	Fewer operations
Datapath precision	Lower precision required	More precision required

Let us now discuss the stability and structure of IIR filters. The transfer function of an IIR filter is expressed as

$$H(z) = \frac{b_0 + b_1 z^{-1} + \ldots + b_N z^{-N}}{1 + a_1 z^{-1} + \ldots + a_M z^{-M}} \qquad (4.3)$$

It is well known that as far as stability is concerned, the direct-form implementation is sensitive to coefficient quantization errors. Noting that the second-order cascade form produces a more robust response to quantization noise [2], the preceding transfer function can be rewritten as

$$H(z) = \prod_{k=1}^{N_s} \frac{b_{0k} + b_{1k} z^{-1} + b_{2k} z^{-2}}{1 + a_{1k} z^{-1} + a_{2k} z^{-2}} \qquad (4.4)$$

where $N_s = \lfloor N/2 \rfloor$, $\lfloor . \rfloor$ represents the largest integer less than or equal to the inside value. This serial or cascaded structure is illustrated in Figure 4-3.

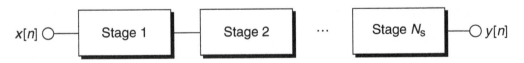

Figure 4-3: Cascaded filter stages.

It is worth mentioning that each second-order filter is considered to be of direct-form II, as shown in Figure 4-4, in order to have a more memory efficient implementation.

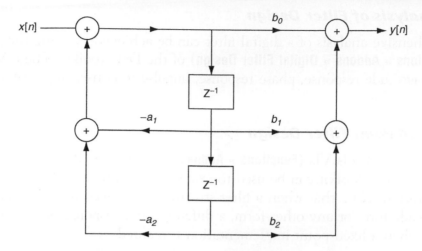

Figure 4-4: Second order direct-form II.

4.2 LabVIEW Digital Filter Design Toolkit

There exist various software tools for designing digital filters. Here, we have used the LabVIEW Digital Filter Design (DFD) toolkit. Any other filter design tool may be used to obtain the coefficients of a desired digital filter. The DFD toolkit provides various tools to design, analyze, and simulate floating-point and fixed-point implementations of digital filters [1].

4.2.1 Filter Design

The Filter Design VIs of the DFD toolkit allow one to design a digital filter with ease by specifying its specifications. For example, the DFD Classical Filter Design Express VI (**Functions » Addons » Digital Filter Design » Filter Design**) provides a graphical user interface to design and analyze digital filters, and the DFD Pole-Zero Placement Express VI (**Functions » Addons » Digital Filter Design » Filter Design**) can be used to alter the locations of poles and zeros in the complex plane.

The filter design methods provided in the DFD toolkits include Kaiser window, Dolph-Chebyshev window, and equi-ripple for FIR filters; and Butterworth, Chebyshev, Inverse Chebyshev, and Elliptic for IIR filters.

In addition, the DFD toolkit has some Special Filter Design VIs. These VIs are used to design special filters such as notch/peak filter, comb filter, maximally flat filter, narrowband filter, and group delay compensator.

4.2.2 Analysis of Filter Design

A comprehensive analysis of a digital filter can be achieved by using the Analysis VIs (**Functions » Addons » Digital Filter Design**) of the DFD toolkit. These VIs provide magnitude response, phase response, impulse response, step response, and zero/pole plot.

4.2.3 Fixed-Point Filter Design

The Fixed-Point Tools VIs (**Functions » Addons » Digital Filter Design » Fixed-Point Tools**) of the DFD toolkit can be used to examine the outcome of a fixed-point implementation. Note that when a filter structure is changed from the direct form to the cascade form or any other form, a different filter response is obtained, in particular when a fixed-point implementation is realized.

4.2.4 Multi-rate Digital Filter Design

The DFD toolkit also provides a group of VIs, named Multirate Filter Design VIs (**Functions » Addons » Digital Filter Design**), for the design, analysis, and implementation of multi-rate filters. These multi-rate filters include single-stage, multi-stage, halfband, Nyquist, raised cosine, and cascaded integrator comb (CIC) filters [1].

4.3 Bibliography

[1] National Instruments, *Digital Filter Design Toolkit User Manual*, Part Number 371353A-01, 2005.

[2] J. Proakis and D. Manolakis, *Digital Signal Processing: Principles, Algorithms, and Applications*, Prentice-Hall, 1995.

Lab 4: FIR/IIR Filtering System Design

In this lab, an FIR and an IIR filter are designed using the VIs as part of the Digital Filter Design (DFD) toolkit. In addition, a point-by-point FIR filter system in hybrid mode is implemented utilizing the Call Library Function feature of LabVIEW.

L4.1 FIR Filtering System

An FIR lowpass filtering system is designed and built in this section.

L4.1.1 Design FIR Filter with DFD Toolkit

Let us design a lowpass filter having the following specifications: passband response = 0.1 dB, passband frequency = 1200 Hz, stopband attenuation = 30 dB, stopband frequency = 2200 Hz, and sampling rate = 8000 Hz. In order to design this filter using the DFD toolkit, place the DFD Classical Filter Design Express VI (**Functions » Addons » Digital Filter Design » Filter Design » DFD Classical Filter Design**) on the BD. Enter the specifications of the filter in the configuration dialog box which appears when placing this Express VI. The magnitude response of the filter and the zero/pole plot are displayed based on the filter specifications in the configuration dialog box, as shown in Figure L4-1. Here, the equi-ripple method is chosen as the design method.

Once this Express VI is configured, its label is changed based on the filter type specified, e.g., Equi-Ripple FIR Lowpass Filter in this example. The filter type gets displayed on the BD, as shown in Figure L4-2.

Additional information on the designed filter such as phase, group delay, impulse response, unit response, frequency response, and zero/pole plot can be seen by using the DFD Filter Analysis Express VI (**Functions » Addons » Digital Filter Design » Filter Analysis » DFD Filter Analysis**). As indicated in Figure L4-2, wire five waveform graphs to the output terminals of the DFD Filter Analysis Express VI except for the Z Plane terminal. The DFD Pole-Zero Plot control (**Controls » Addons » Digital Filter Design » DFD Pole-Zero Plot**) needs to be placed on the FP to obtain the zero/pole plot. This locates a terminal icon on the BD. Then, wire the Z Plane terminal of the DFD Filter Analysis Express VI to the DFD Pole-Zero Plot control.

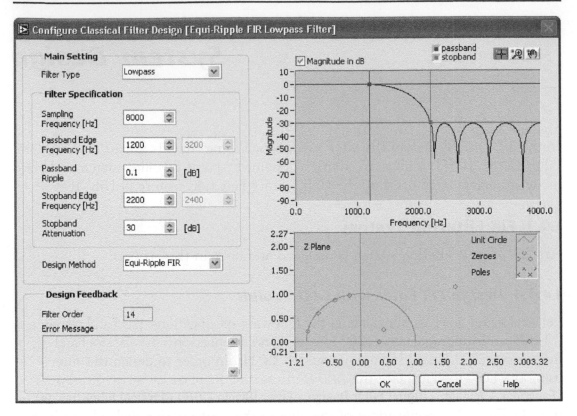

Figure L4-1: Configuration of FIR lowpass filter.

The coefficients of the filter are obtained by wiring the DFD Get TF VI
(**Functions » Addons » Digital Filter Design » Utilities » DFD Get TF**) to the filter
cluster, i.e., the output of the DFD Classical Filter Design Express VI.
The DFD Get TF VI retrieves the transfer function of the filter designed by the
DFD Classical Filter Design Express VI. For FIR filters, the numerator
values of the transfer function correspond to the b coefficients of the filter and the
denominator to unity. The transfer function of the designed filter can be observed by
creating two Numeric Indicators. To do this, right-click on the numerator
terminal of the DFD Get TF VI and choose **Create » Indicator** from the shortcut
menu. The second indicator is created and wired to the denominator terminal
of the VI.

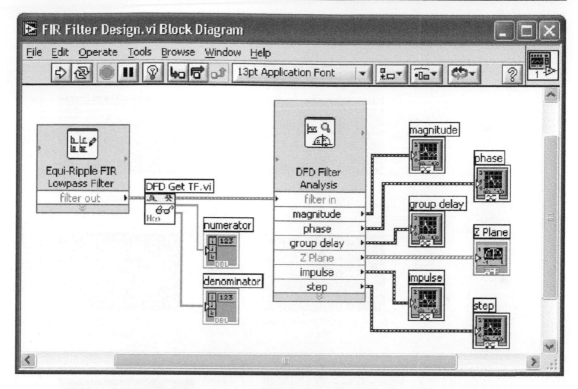

Figure L4-2: Design and analysis of FIR filter using the DFD toolkit.

Save the VI as *FIR Filter Design.vi* and then run it. The response of the designed FIR filter is illustrated in Figure L4-3. Notice that the indicator array on the FP needs to be resized to display all the elements of the coefficient set.

L4.1.2 Creating a Filtering System VI

The VI of the filtering system built here consists of signal generation, filtering, and graphical output components. Three sinusoidal signals are summed and passed through the designed FIR filter and the filtered signal is then displayed and verified.

Let us build the FP of the filtering system. Place three Horizontal Pointer Slide controls (**Controls » Modern » Numeric » Horizontal Pointer Slide**) to adjust the frequency of the signals. Place three waveform graphs to display the input signal and

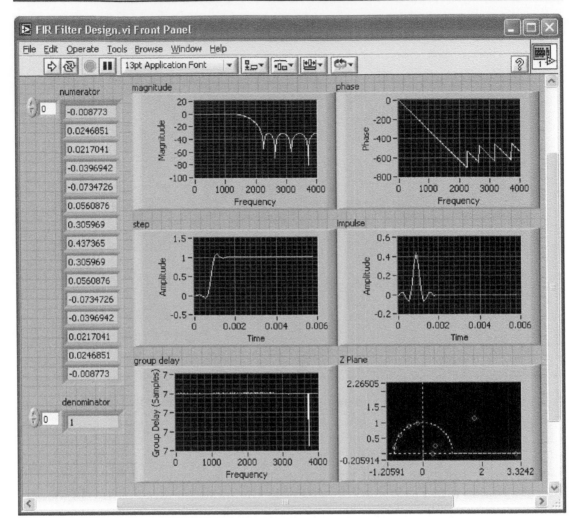

Figure L4-3: FP of FIR filter object.

filtered signal in the time and frequency domains. Observe that the corresponding terminal icons for the Horizontal Pointer Slide controls and waveform graphs get created on the BD, as shown in Figure L4-4.

Next, switch to the BD. To provide the signal source of the system, place three Sine Waveform VIs (**Functions » Signal Processing » Waveform Generation » Sine Waveform**) on the BD. The amplitude of the output sinusoid is configured to be its

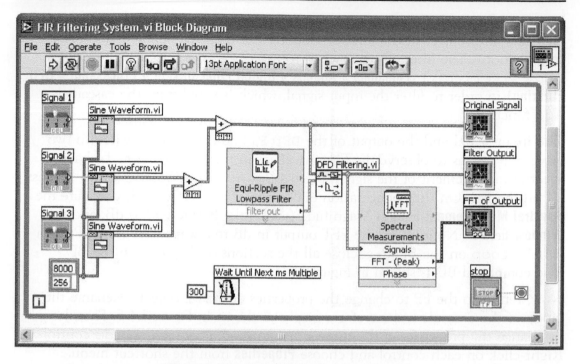

Figure L4-4: BD of FIR filtering system.

default value of unity in the absence of an input. The icons of the Horizontal Pointer Slide controls are wired to the frequency terminal of each Sine Waveform VI.

Create a cluster constant to incorporate the sampling information. This is done by right-clicking on the sampling info terminal of the Sine Waveform VI and choosing **Create » Constant**. Enter 8000 as the sampling rate and 256 as the number of samples. Wire the cluster constant to all three VIs so that all the signals have the same sampling rate and length. The three signal arrays are summed together to construct the input signal of the filtering system. This is done by using two Add functions (**Functions » Programming » Numeric » Add**), as shown in Figure L4-4.

Now the filtering component is described. The filter is designed by using the DFD Classical Filter Design Express VI (**Functions » Addons » Digital Filter**

Design » Filter Design » DFD Classical Filter Design) as described earlier. This VI creates filter object in the form of a cluster based on the configured filter specifications. The filter object is wired to the filter in terminal of the DFD Filtering VI (**Functions » Addons » Digital Filter Design » Processing » DFD Filtering**) in order to filter the input signal, which is wired from the cascaded Add function.

The input signal and the output of the DFD Filtering VI are wired to two waveform graphs to observe the filtering effect in the time domain. To have a spectral measurement of the signal, place a Spectral Measurements Express VI on the BD. On the configuration dialog box of the Express VI, configure the **Spectral Measurement** field as Magnitude (peak), the **Result** field as dB, and the **Window** field as None. Wire the FFT output in dB to a waveform graph. Place a While Loop on the BD to enclose all the sections of the code on the BD. The completed BD is shown in Figure L4-4.

Now, return to the FP to change the properties of the FP objects. Rename the labels of the controls and waveform graphs as shown in Figure L4-5. First, let us change the properties of the three Horizontal Pointer Slide controls. Right-click on each control and choose **Properties** from the shortcut menu. This brings up a properties dialog box. Change the maximum scale of all the three controls to the Nyquist frequency, 4000 Hz, in the **Scale** tab, and set the default frequency values to 750 Hz, 2500 Hz, and 3000 Hz, respectively, in the **Data Range** tab.

Next, let us modify the properties of the waveform graph labeled as FFT of Output in Figure L4-5. Right-click on the waveform graph and choose **Properties** from the shortcut menu to bring up a properties dialog box. Uncheck **Autoscale** of the Y axis and change the minimum scale to −80 in the **Scales** tab to observe peaks of the waveform more closely. In the other two graphs corresponding to the time domain signal, uncheck **X Scale » Loose Fit** from the shortcut menu to fit the plot into the entire plotting area.

Save the VI as *FIR Filtering System.vi* and then run it. Note that among the three signals 750 Hz, 2500 Hz, and 3000 Hz, the 2500 Hz and 3000 Hz signals should be filtered out and only the 750 Hz signal should be seen at the output. The waveform result on the FP during run time is shown in Figure L4-5.

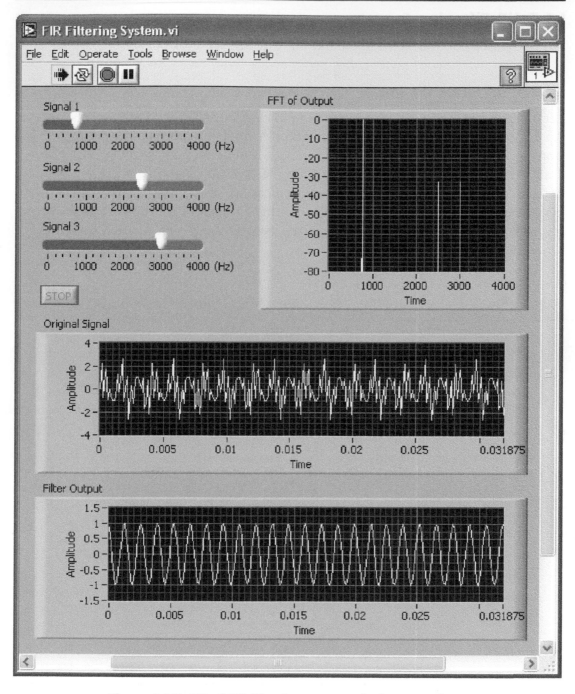

Figure L4-5: FP of FIR filtering system during run time.

L4.2 IIR Filtering System

An IIR bandpass filter is designed and built in this section.

L4.2.1 IIR Filter Design

Let us consider a bandpass filter with the following specifications: passband response = 0.5 dB, passband frequency = 1333 to 2666 Hz, stopband attenuation = 20 dB, stopband frequency = 1000 to 3000 Hz, and sampling frequency = 8000 Hz. The design of an IIR filter is achieved by using the DFD Classical Filter Design Express VI described earlier. Enter the specifications of the filter in the configuration dialog box which is brought up by placing this Express VI on the BD, as shown in Figure L4-6. The elliptic method is chosen here as the design method to achieve a narrow transition band.

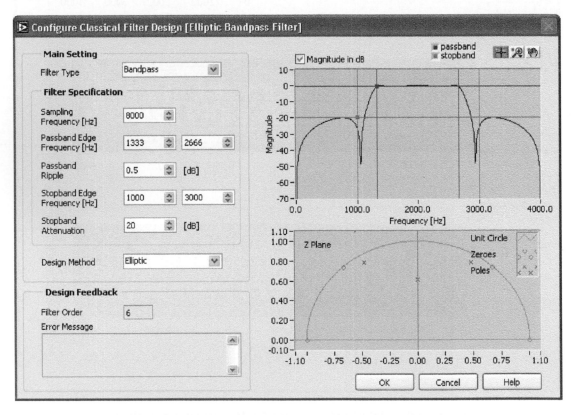

Figure L4-6: Configuration of IIR bandpass filter.

Similar to FIR filtering, the label of the Express VI is changed to `Elliptic Bandpass Filter` by altering the configuration as shown in Figure L4-7. The response of the designed filter can be obtained by using the `DFD Filter Analysis` Express VI and wiring five waveform graphs and a `DFD Pole-Zero Plot` control. These steps are similar to those mentioned for FIR filtering.

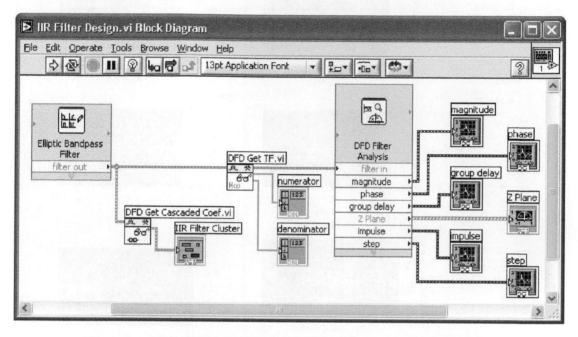

Figure L4-7: Design and analysis of IIR filter using DFD toolkit.

The filter coefficients provided by the `DFD Classical Filter Design` Express VI correspond to the "IIR cascaded second order sections form II" structure by default. To observe the cascaded coefficients, one can wire the filter cluster to the `DFD Get Cascaded Coef` VI. A cluster of indicators is created by right-clicking on the `IIR Filter Cluster` terminal of the VI and choosing **Create » Indicator**. The filter coefficients corresponding to the "IIR direct form II" structure are obtained by using the `DFD Get TF` VI similar to FIR filtering.

Save the VI as *IIR Filtering Design.vi* and then run it. The response of the IIR bandpass filter is illustrated in Figure L4-8.

Notice that the filter coefficients are displayed as truncated values in Figure L4-8. The format of the numeric indicators is configured to be floating-point with 6 digits of precision. This is done by right-clicking on the numeric indicators

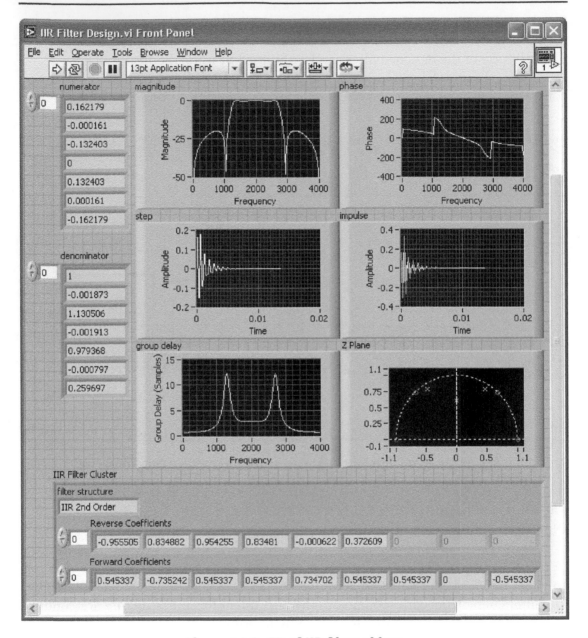

Figure L4-8: FP of IIR filter object.

on the FP and choosing **Format & Precision** ... from the shortcut menu.
A properties dialog box is brought up, as shown in Figure L4-9. Configure the
representation to **Floating point** and the precision to **6 Digits of precision,** as shown
in Figure L4-9.

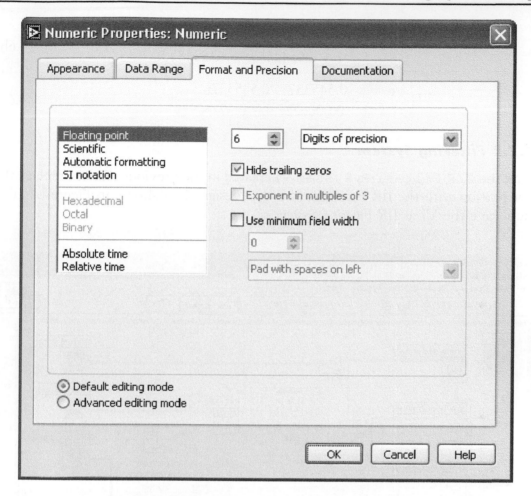

Figure L4-9: Changing properties of numeric indicator format and precision.

From the coefficient set, the transfer function of the IIR filter is given by

$$H[z] = \frac{0.162179 - 0.000161z^{-1} - 0.132403z^{-2} + 0.132403z^{-4} + 0.000161z^{-5} - 0.162179z^{-6}}{1 - 0.001873z^{-1} + 1.130506z^{-2} - 0.001913z^{-3} + 0.979368z^{-4} - 0.000797z^{-5} + 0.259697z^{-6}}$$

$$= H_1[z] \cdot H_2[z] \cdot H_3[z] \tag{4.5}$$

where, $H_1[z]$, $H_2[z]$, and $H_3[z]$ denote the transfer functions of the three second-order sections. From the cascade coefficient cluster, the three transfer functions are

$$H_1[z] = \frac{0.545337 - 0.735242z^{-1} + 0.545337z^{-2}}{1 - 0.955505z^{-1} + 0.834882z^{-2}} \tag{4.6a}$$

$$H_2[z] = \frac{0.545337 + 0.734702z^{-1} + 0.545337z^{-2}}{1 + 0.954255z^{-1} + 0.834810z^{-2}} \qquad (4.6b)$$

$$H_3[z] = \frac{0.545337 - 0.545337z^{-2}}{1 - 0.000622z^{-1} + 0.372609z^{-2}} \qquad (4.6c)$$

L4.2.2 Filtering System

Using the FIR Filtering System VI created in the previous section, replace the filter portion with the IIR bandpass filter just designed, as shown in Figure L4-10. Then, save the VI as *IIR Filtering System.vi*.

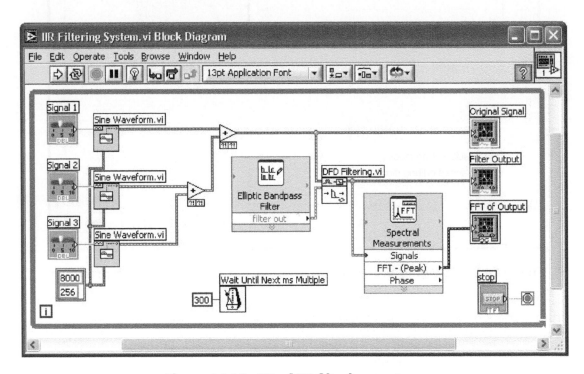

Figure L4-10: BD of IIR filtering system.

Let us change the default values of the three frequency controls on the FP to 1000 Hz, 2000 Hz, and 3000 Hz to see whether the IIR filter is functioning properly. The signals having the frequencies 1000 Hz and 3000 Hz should be filtered out while only the signal having the frequency 2000 Hz should remain and be seen in the output. The output waveform as seen on the FP is shown in Figure L4-11. From the FFT of the output, one can see that the desired stopband attenuation of 20 dB is obtained.

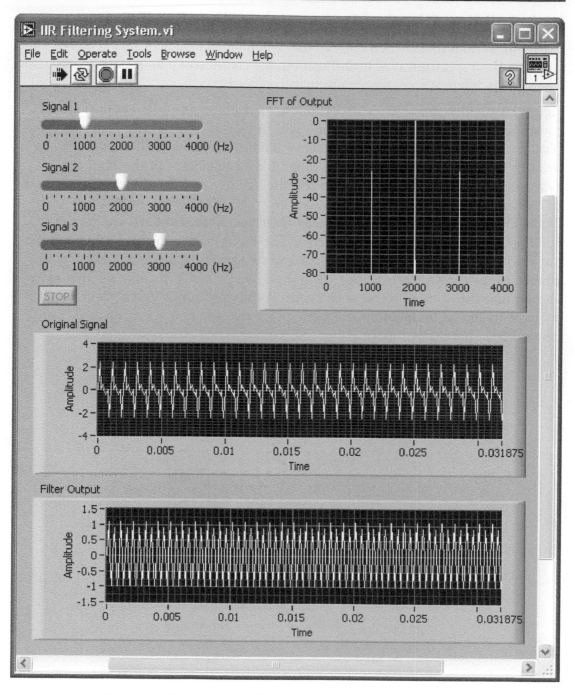

Figure L4-11: FP of IIR filtering system during run time.

L4.3 Building Filtering System Using Filter Coefficients

There are various tools which one can use to compute coefficient sets of digital filters based on their specifications. In this section, the creation of a filter object is discussed when using different tools for obtaining its coefficients.

Figure L4-12 illustrates two ways to build a filter object using arrays of numeric constants containing filter coefficients. The DFD Build Filter from TF VI (**Functions » Addons » Digital Filter Design » Utilities » DFD Build Filter from TF**) can be used to build a filter object if the direct-form coefficients of the filter are available; see Figure L4-12(a). For an IIR filter in the second-order cascade form, the DFD Build Filter from Cascaded Coef VI (**Functions » Addons » Digital Filter Design » Utilities » DFD Build Filter from Cascaded Coef**) can be used to build a filter object; see Figure L4-12(b). The input cluster to this VI consists of a numeric constant for the filter structure and two arrays of numeric constants, labeled as Reverse Coefficients and Forward Coefficients. Each filter section consists of two reverse coefficients in the denominator, and three forward coefficients in the numerator, considering that the first coefficient of the denominator is regarded as 1.

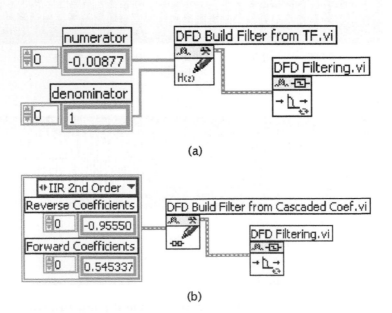

(a)

(b)

Figure L4-12: DFD Build Filter: (a) using direct-form coefficients and (b) using cascade-form coefficients.

L4.4 Filter Design Without Using DFD Toolkit

The examples explained in the preceding sections can be implemented without using the DFD toolkit. One can achieve this by using the `Digital FIR Filter` VI (**Functions » Signal Processing » Waveform Conditioning » Digital FIR Filter**) and `Digital IIR Filter` VI (**Functions » Signal Processing » Waveform Conditioning » Digital IIR Filter**).

Similar to the `Classic Filter Design` Express VI of the DFD toolkit, the `Digital FIR Filter` VI is configured based on the filter specifications; thus, one does not need to obtain the filter coefficients before building the filtering system. As a result, the specifications can be adjusted on the fly. The BD corresponding to this approach is shown in Figure L4-13.

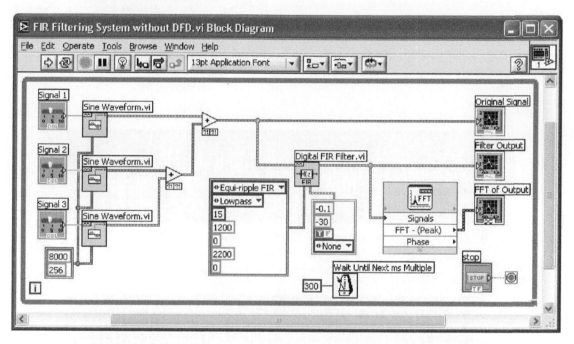

Figure L4-13: BD of FIR filtering system without using DFD VI.

For the `Digital FIR Filter` VI, the filter specifications are defined via two inputs in the form of a cluster constant. A cluster constant is created by right-clicking on the `FIR filter specifications` terminal and choosing **Create » Constant**. This cluster specifies the filter type, number of taps, and lower/upper passband or stopband. Another cluster constant specifying the passband gain, stopband gain, and window type can be wired to the `Optional FIR filter specifications` terminal. More details on the use of cluster constants as related to the `Digital FIR Filter` VI can be found in [1].

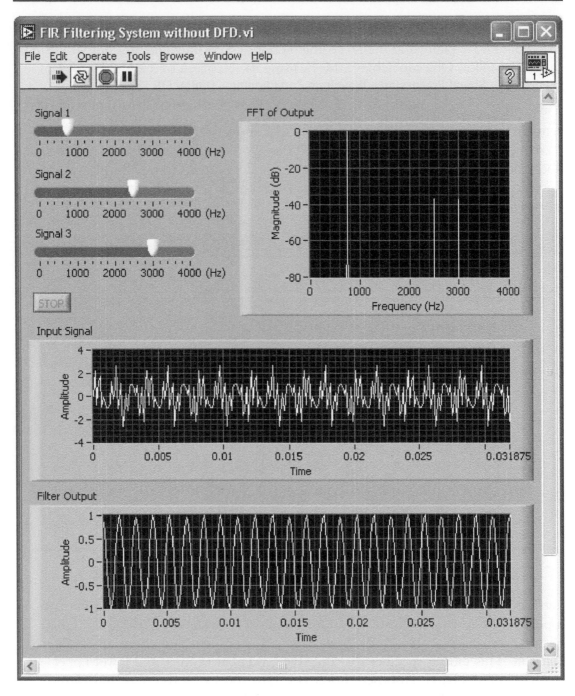

Figure L4-14: FP of filtering system using specifications.

Rename the FP objects and set the maximum and default values of the controls. Save the VI as *FIR Filtering System without DFD.vi* and then run it. The FP of the VI during run time is shown in Figure L4-14. Observe that the 750 Hz signal falling in the passband remains, while the 2500 Hz and 3000 Hz signals falling in the stopband, i.e., greater than 2200 Hz, are attenuated by 30 dB.

L4.5 Building Filtering System Using Dynamic Link Library (DLL)

When a system is being implemented, it is often more efficient to merge existing modules from available or previously written codes. In this section, an FIR filter is implemented as a C DLL and called in a point-by-point fashion within the LabVIEW programming environment.

L4.5.1 Point-by-Point Processing

Before one builds a filtering system using DLL, it is useful to become familiar with the point-by-point processing feature of LabVIEW. As implied by its name, point-by-point processing is a scalar type of data processing. Point-by-point processing is suitable for real-time data processing tasks, such as signal filtering, since it allows inputs and outputs to be synchronized. On the other hand, in array-based processing, there exists a delay between data acquisition and processing [2].

Figure L4-15 shows the BD of an FIR filtering system utilizing point-by-point processing. A single sample of the input signal, which consists of three sinusoids, gets generated at each iteration of the While Loop by using the Sine Wave PtByPt VI (**Functions » Signal Processing » Point By Point » Signal Generation PtByPt » Sine Wave PtByPt**). This VI requires a normalized frequency input. Thus, the signal frequency is divided by the sampling frequency, 8000 Hz, and is wired to the f terminal of the VI. Also, the iteration counter of the While Loop is wired to the time terminal of the VI.

The FIR Filter PtByPt VI (**Functions » Signal Processing » Point By Point » Filters PtByPt » FIR Filter PtByPt**) is used here to serve as the FIR filter. The filter coefficients obtained from the DFD toolkit (section L4.1.1) are entered and wired to the VI. This is done by right-clicking on the forward coefficients terminal and choosing **Create » Constant** from the shortcut.

The filtered output signal is then examined in both the time and frequency domains. The FFT PtByPt VI (**Functions » Signal Processing » Point By Point » Transforms PtByPt » FFT PtByPt**) is used to see the frequency response. Note that this VI collects a frame of the incoming samples to compute the FFT. A total of

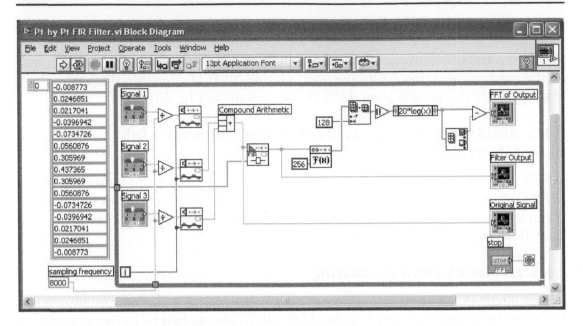

Figure L4-15: BD of point-by-point FIR filtering system.

256 input samples is used to generate the magnitudes of 256 complex FFT values. Only the first half of the values is displayed in normalized magnitude, considering that the second half is a mirror image of the first half. This is done by using the Array Subset function. Notice that the index is set to its default value, i.e., 0, in the absence of an input to ensure the consistency of the magnitude spectrum display. The FFT outcome is converted to dB and normalized by its maximum and then displayed in a waveform graph.

For the time domain observation of the signals, waveform charts (**Controls » Modern » Graph » Waveform Chart**) are used instead of waveform graphs. Waveform charts are further discussed in Chapter 6.

The FP of the FIR filtering system is shown in Figure L4-16. To modify the display length, right-click on the plot area of the waveform chart and choose **Chart History Length** .... This brings up a dialog box to adjust the number of samples for the display. Enter 256 for the buffer length.

Let us change the properties of the FP objects. Rename the axes of the waveform graphs as shown in Figure L4-16. The scale factor needs to be modified in order to have a proper scaling of the frequency axis on the waveform graphs. The value of 4000/128 = 31.25 is used as the multiplier of the X axis to scale it in the range 0 to π (radians), that is, 4000 Hz. This is done by right-clicking on

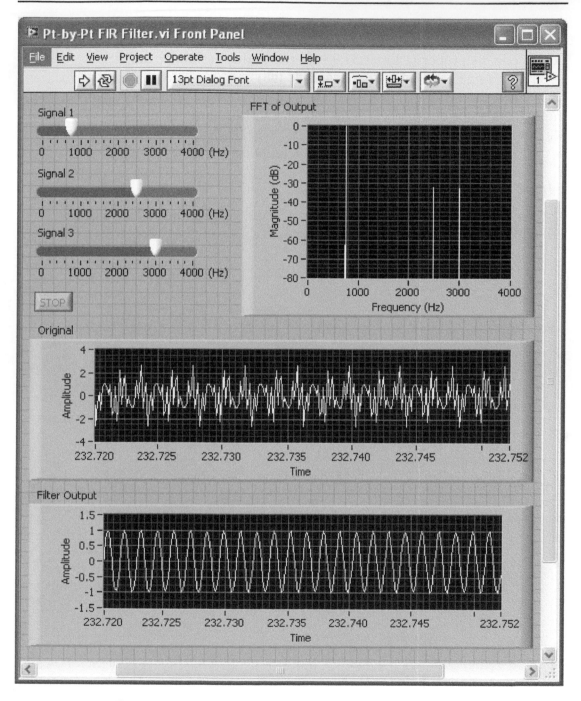

Figure L4-16: FP of point-by-point FIR filtering system.

the waveform graphs and choosing **Properties**. This brings up the Waveform Graph Property window. Click the **Scales** tab and choose the **Frequency (Hz)** axis to edit its property. Enter 31.25 for the multiplier under the Scaling factors field. The **Magnitude (dB)** axis properties need to be edited as well. To keep the magnitude spectrum display consistent, disable the **Autoscale** option for the Y axis and set the Minimum field and Maximum field to –80 and 0, respectively. The FP should display the outcome as that of Figure L4-16 when all the axis properties are set correctly.

L4.5.2 Creating DLL in C

Now let us implement the FIR filter in C and build a DLL that can be called by LabVIEW. Consider the following filtering source code in C for generating a DLL file:

```
#include <windows.h>
#include <string.h>
#include <ctype.h>

BOOL WINAPI DllMain (HANDLE hModule, DWORD dwFunction, LPVOID lpNot)
{
    return TRUE;
}
/* This function implements FIR filter */
_declspec (dllexport) double FIR(double input, double *
inputBuffer)
{
    int i;
    int bufferLength=15;
    double h[15]={-0.008773, 0.0246851, 0.0217041, -0.0396942,
-0.0734726, 0.0560876,
        0.305969, 0.437365,0.305969, 0.0560876, -0.0734726,
-0.0396942, 0.0217041, 0.0246851, -0.008773};
    double sum=0;

    // shift data in the input buffer
    for(i=bufferLength-1; i>0; i--)
    {
            inputBuffer[i]=inputBuffer[i-1];
    }
    inputBuffer[0]=input;
    // calculate output
    for(i=0; i<bufferLength; i++)
    sum=sum+inputBuffer[i]*h[i];

    return sum;
}
```

To generate the DLL file, create a new Win32 project in Microsoft Visual Studio and include the preceding C code as the source file. After the project is built, the DLL file can be seen in the debug folder.

L4.5.3 Calling DLL from LabVIEW

With little modification, the point-by-point FIR filter system built earlier can be turned into a hybrid system incorporating the DLL file. As shown in Figure L4-17, the `FIR Filter PtByPt` VI is replaced by a `Call Library Function Node` VI (**Connectivity » Libraries & Executables » Call Library Function Node**). To call the DLL, one has to properly configure this VI. Double-click on the VI to bring up a dialog box for configuring the function node. Under **Function** tab, **Library name or path** must be specified as the path of the DLL file. **Function name** must be set to have the same name as the one defined in the source code (FIR in this case). Also, select **Run in UI thread** for **Thread** and **C** for **Calling convention**. The parameters of the function node must be added and configured under the **Parameters** tab. The number and data type of parameters must match those of the function defined in the source code. The first parameter (`output`) is return type, and the rest (`input` and **inputBuffer**) are function inputs. For the `input` and `output` parameters, **Type** is set to **Numeric, Data type** is set to **8-byte Double**, and **Pass** is

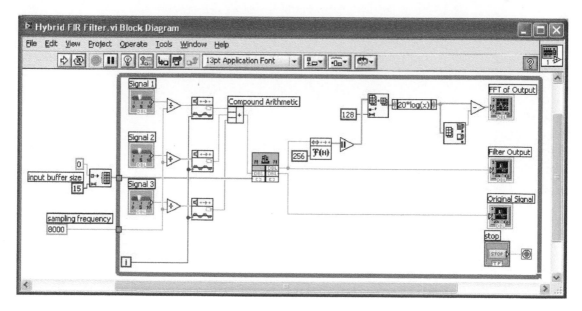

Figure L4-17: BD of FIR filtering system using DLL.

set to **Value**. For the `inputBuffer` parameter, **Type** is set to **Array, Data type** is set to **8-byte Double, Dimensions** is set to **1**, and **Array format** is set to **Array Data Pointer**.

Notice that an array is initialized outside the `While Loop` and wired to the `inputBuffer` terminal of the function node. This way, memory is allocated to store input samples, with the memory address getting passed into the function node. The array size is set to 15, the length of the filter. This FIR filter system built in hybrid mode produces the same output as the system appearing in Figure L4-16.

L4.6 Bibliography

[1] National Instruments, *Signal Processing Toolset User Manual*, Part Number 322142C-01, 2002.

[2] National Instruments, *LabVIEW Analysis Concepts*, Part Number 370192C-01, 2004.

L4.7 Lab Experiments

Perform the first two experiments by using the MathScript feature and the next two experiments without using the MathScript feature.

1. Build a VI to eliminate the frequency component f_2 Hz from the composite signal given by Equation (4.7). The composite signal in Equation (4.7) consists of three sinusoids with frequencies f_1 Hz, f_2 Hz, f_3 Hz, and amplitudes A_1, A_2, A_3, respectively. It is based on a sampling frequency of 8500 Hz with the number of samples being 400. Set the f_1 Hz, f_2 Hz, and f_3 Hz frequency ranges to 20 Hz – 60 Hz, 150 Hz – 240 Hz, and 700 Hz – 900 Hz, respectively. Set the A_1, A_2, and A_3 amplitude ranges to 2–5, 7–10, and 12–15, respectively. Generate the composite signal and display it together with the filtered signal. Also, display the FFT spectrum of the composite signal and the filtered signal.

$$x(t) = A_1^{(3/2)} \sin\left(2\pi f_1 t\right) + A_2^{(1/2)} \cos\left(2\pi f_2 t\right) + A_3 \sin\left(2\pi f_3 t\right) \tag{4.7}$$

2. Build a VI to decompose the composite signal given by Equation (4.8) into its individual frequency components. The composite signal in Equation (4.8) consists of three sinusoids with frequencies f_1 Hz, f_2 Hz, f_3 Hz, and amplitudes A_1, A_2, A_3, respectively. It is based on a sampling frequency of 7500 Hz with the number of samples being 400. Set the f_1 Hz, f_2 Hz, and f_3 Hz frequency ranges to 40 Hz – 80 Hz, 250 Hz – 360 Hz, and 850 Hz – 1000 Hz, respectively. Set the A_1, A_2, and A_3 amplitude ranges from 3–8, 14–20, and 22–35, respectively. Generate the composite signal and display it together with the filtered signals. Also, display the FFT spectrum of the composite signal and the filtered signals.

$$x(t) = A_1^{(5/2)} \sin\left(2\pi f_1 t\right) + A_2^{(3/2)} \cos\left(2\pi f_2 t\right) + A_3 \sin\left(2\pi f_3 t\right) \tag{4.8}$$

3. Build a VI to estimate the order of the Butterworth IIR filter with a sampling frequency of 8000 Hz based on the frequency, ripple, and filter type specifications. Display the estimated filter order, the cascaded filter coefficients together with the magnitude and phase response of the filter. Also, find the inverse filter and check its stability. Issue a warning message if it becomes unstable.

4. Build a VI to estimate the order of the Chebyshev IIR filter with a sampling frequency of 8000 Hz based on the frequency, ripple, and filter type specifications. Display the estimated filter order, the cascaded filter coefficients together with the magnitude and phase response of the filter. Also, find the inverse filter and check its stability. Issue a warning message if it becomes unstable.

Fixed-Point versus Floating-Point

From an arithmetic point of view, there are two ways a DSP system can be implemented in LabVIEW to match its hardware implementation on a processor. These include fixed-point and floating-point implementations. In this chapter, we discuss the issues related to these two hardware implementations.

In a fixed-point processor, numbers are represented and manipulated in integer format. In a floating-point processor, in addition to integer arithmetic, floating-point arithmetic can be handled. This means that numbers are represented by the combination of a mantissa (or a fractional part) and an exponent part, and the processor possesses the necessary hardware for manipulating both of these parts. As a result, in general, floating-point processors are slower than fixed-point ones.

In a fixed-point processor, one needs to be concerned with the dynamic range of numbers, since a much narrower range of numbers can be represented in integer format as compared to floating-point format. For most applications, such a concern can be virtually ignored when using a floating-point processor. Consequently, fixed-point processors usually demand more coding effort than do their floating-point counterparts.

5.1 Q-format Number Representation

The decimal value of an N-bit 2's-complement number, $B = b_{N-1}b_{N-2}...b_1b_0, b_i \in \{0,1\}$, is given by

$$D(B) = -b_{N-1}2^{N-1} + b_{N-2}2^{N-2} + ... + b_1 2^1 + b_0 2^0 \qquad (5.1)$$

The 2's-complement representation allows a processor to perform integer addition and subtraction by using the same hardware. When the unsigned integer representation is used, the sign bit is treated as an extra bit. This way, only positive numbers can be represented.

There is a limitation of the dynamic range of the foregoing integer representation scheme. For example, in a 16-bit system, it is not possible to represent numbers larger than $2^{15} - 1 = 32767$ and smaller than $-2^{15} = -32768$. To cope with this limitation, numbers are often normalized between -1 and 1. In other words, they are represented as fractions. This normalization is achieved by the programmer moving the implied or imaginary binary point (note that there is no physical memory allocated to this point) as indicated in Figure 5-1. This way, the fractional value is given by

$$F(B) = -b_{N-1}2^0 + b_{N-2}2^{-1} + \ldots + b_1 2^{-(N-2)} + b_0 2^{-(N-1)} \tag{5.2}$$

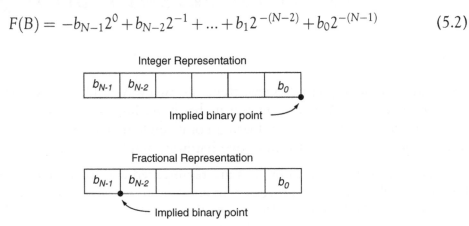

Figure 5-1: Number representations.

This representation scheme is referred to as the Q-format or fractional representation. The programmer needs to keep track of the implied binary point when manipulating Q-format numbers. For instance, let us consider two Q15 format numbers. Each number consists of 1 sign bit plus 15 fractional bits. When these numbers are multiplied, a Q30 format number is generated (the product of two fractions is still a fraction), with bit 31 being the sign bit and bit 32 another sign bit (called an extended sign bit). Assuming a 16-bit wide memory, not enough bits are available to store all 32 bits, and only 16 bits can be stored. It makes sense to store the 16 most significant bits. This requires storing the upper portion of the 32-bit product by doing a 15-bit right shift. In this manner, the product would be stored in Q15 format. (See Figure 5-2.)

Based on the 2's-complement representation, a dynamic range of $-2^{N-1} \leq D(B) \leq 2^{N-1} - 1$ can be covered, where N denotes the number of bits. For illustration purposes, let us consider a 4-bit system where the most negative number is -8 and the most positive number is 7. The decimal representations of the numbers are shown in Figure 5-3. Notice how the numbers change from most positive to most negative with the sign bit. Since only the integer numbers falling within

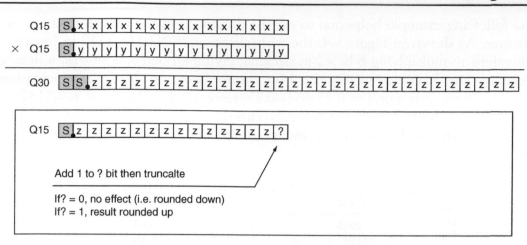

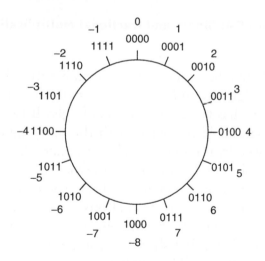

Q15 $\boxed{S}$ z z z z z z z z z z z z z z z $?$

Add 1 to ? bit then truncalte

If? = 0, no effect (i.e. rounded down)
If? = 1, result rounded up

Figure 5-2: Multiplying and storing Q15 numbers.

the limits –8 and 7 can be represented, one can easily see that any multiplication or addition resulting in a number larger than 7 or smaller than –8 will cause overflow. For example, when 6 is multiplied by 2, the result is 12. Hence, the result is greater than the representation limits and will be wrapped around the circle to 1100, which is –4.

Figure 5-3: Four-bit binary representation.

The Q-format representation solves this problem by normalizing the dynamic range between –1 and 1. This way, any resulting multiplication will be within these limits. Using the Q-format representation, the dynamic range is divided into 2^N sections, where $2^{-(N-1)}$ is the size of a section. The most negative number is always –1 and the most positive number is $1-2^{-(N-1)}$.

125

The following example helps one to see the difference in the two representation schemes. As shown in Figure 5-4, the multiplication of 0110 by 1110 in binary is equivalent to multiplying 6 by –2 in decimal, giving an outcome of –12, a number exceeding the dynamic range of the 4-bit system. Based on the Q3 representation, these numbers correspond to 0.75 and –0.25, respectively. The result is –0.1875, which falls within the fractional range. Notice that the hardware generates the same 1's and 0's; what is different is the interpretation of the bits.

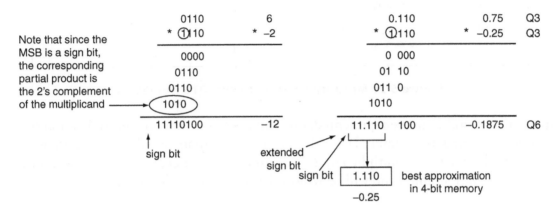

Figure 5-4: Binary and fractional multiplication.

When multiplying Q-N numbers, one should remember that the result will consist of 2N fractional bits, one sign bit, and one or more extended sign bits. Based on the data type used, the result has to be shifted accordingly. If two Q15 numbers are multiplied, the result will be 32 bits wide, with the most significant bit being the extended sign bit followed by the sign bit. The imaginary decimal point will be after the 30th bit. So a right shift of 15 is required to store the result in a 16-bit memory location as a Q15 number. It should be realized that some precision is lost, of course, as a result of discarding the smaller fractional bits. Since only 16 bits can be stored, the shifting allows one to retain the higher precision fractional bits. If a 32-bit storage capability is available, a left shift of 1 can be done to remove the extended sign bit and store the result as a Q31 number.

To further understand a possible precision loss when manipulating Q-format numbers, let us consider another example where two Q12 numbers corresponding to 7.5 and 7.25 are multiplied. As can be seen from Figure 5-5, the resulting product must be left-shifted by 4 bits to store all the fractional bits corresponding to Q12 format.

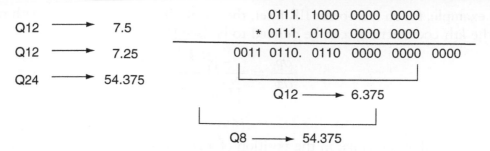

Figure 5-5: Q-format precision loss example.

However, doing so results in a product value of 6.375, which is different than the correct value of 54.375. If the product is stored in a lower precision Q-format—say, in Q8 format—then the correct product value can be obtained and stored.

Although Q-format solves the problem of overflow during multiplications, addition and subtraction still pose a problem. When two Q15 numbers are being added, the sum could exceed the range of the Q15 representation. Solving this problem requires employing the scaling approach, discussed later in this chapter.

5.2 Finite Word Length Effects

Due to the fact that the memory or registers of a processor have a finite number of bits, there could be a noticeable error between desired and actual outcomes on a fixed-point processor. The so-called finite word length quantization effect is similar to the input data quantization effect introduced by an A/D converter.

Consider fractional numbers quantized by a $b+1$ bit converter. When these numbers are manipulated and stored in an $M+1$ bit memory, with $M < b$, there is going to be an error (simply because $b - M$ of the least significant fractional bits are discarded or truncated). This finite word length error could alter the behavior of a DSP system by an unacceptable degree. The range of the magnitude of truncation error ε_t is given by $0 \le |\varepsilon_t| \le 2^{-M} - 2^{-b}$. The lowest level of truncation error corresponds to the situation when all the thrown-away bits are zeros, and the highest level to the situation when all the thrown-away bits are ones.

This effect has been extensively studied for FIR and IIR filters; for example, see [1]. Since the coefficients of such filters are represented by a finite number of bits, the roots of their transfer function polynomials, or the positions of their zeros and poles, shift in the complex plane. The amount of shift in the positions of poles and zeros can be related to the amount of quantization errors in the coefficients.

For example, for an Nth-order IIR filter, the sensitivity of the ith pole p_i with respect to the kth coefficient a_k can be derived to be (see [1])

$$\frac{\partial p_i}{\partial a_k} = \frac{-p_i^{N-k}}{\displaystyle\prod_{\substack{l=1 \\ l \neq i}}^{N} (p_i - p_l)} \tag{5.3}$$

This means that a change in the position of a pole is influenced by the positions of all the other poles. That is the reason the implementation of an Nth order IIR filter is normally achieved by having a number of second-order IIR filters in cascade or series in order to decouple this dependency of poles.

Also, note that as a result of coefficient quantization, the actual frequency response $\hat{H}(e^{j\theta})$ is different than the desired frequency response $H(e^{j\theta})$. For example, for an FIR filter having N coefficients, it can be easily shown that the amount of error in the magnitude of the frequency response, $|\Delta H(e^{j\theta})|$, is bounded by

$$|\Delta H(e^{j\theta})| = |H(e^{j\theta}) - \hat{H}(e^{j\theta})| \leq N2^{-b} \tag{5.4}$$

In addition to the preceding effects, coefficient quantization can lead to limit cycles. This means that in the absence of an input, the response of a supposedly stable system (poles inside the unit circle) to a unit sample is oscillatory instead of diminishing in magnitude.

5.3 Floating-Point Number Representation

Due to relatively limited dynamic ranges of fixed-point processors, when using such processors, one should be concerned with the scaling issue, or how big the numbers get in the manipulation of a signal. Scaling is not of concern when using floating-point processors, since the floating-point hardware provides a much wider dynamic range.

As an example, let us consider the C67x processor, which is the floating-point version of the TI family of TMS320C6000 DSP processors. There are two floating-point data representations on the C67x processor: single precision (SP) and double precision (DP). In the single-precision format, a value is expressed as (see [2])

$$-1^s \times 2^{(exp-127)} \times 1.frac \tag{5.5}$$

where s denotes the sign bit (bit 31), *exp* denotes the exponent bits (bits 23 through 30), and *frac* denotes the fractional or mantissa bits (bits 0 through 22). (See Figure 5-6.)

31	30		23	22		0
s	exp				frac	

Figure 5-6: C67x floating-point data representation.

Consequently, numbers as big as 3.4×10^{38} and as small as 1.175×10^{-38} can be processed. In the double-precision format, more fractional and exponent bits are used as indicated in

$$-1^s \times 2^{(exp - 1023)} \times 1.frac \tag{5.6}$$

where the exponent bits are from bits 20 through 30, and the fractional bits are all the bits of a first word and bits 0 through 19 of a second word. (See Figure 5-7.) In this manner, numbers as big as 1.7×10^{308} and as small as 2.2×10^{-308} can be handled.

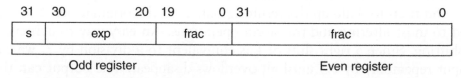

31	30	20	19	0	31	0
s	exp		frac		frac	

Odd register   Even register

Figure 5-7: C67x double-precision floating-point representation.

When one is using a floating-point processor, all the steps needed to perform floating-point arithmetic are done by the floating-point hardware. For example, consider adding two floating-point numbers represented by

$$a = a_{frac} \times 2^{a_{exp}}$$
$$b = b_{frac} \times 2^{b_{exp}} \tag{5.7}$$

The floating-point sum c has the following exponent and fractional parts:

$$
\begin{aligned}
c &= a + b \\
&= \left(a_{frac} + \left(b_{frac} \times 2^{-(a_{exp} - b_{exp})}\right)\right) \times 2^{a_{exp}} && \text{if } a_{exp} \geq b_{exp} \\
&= \left(\left(a_{frac} \times 2^{-(b_{exp} - a_{exp})}\right) + b_{frac}\right) \times 2^{b_{exp}} && \text{if } a_{exp} < b_{exp}
\end{aligned} \tag{5.8}
$$

These parts are computed by the floating-point hardware. This shows that, though possible, it is inefficient to perform floating-point arithmetic on fixed-point processors, since all the operations involved, such as those in Equation (5.8), must be done in software.

5.4 Overflow and Scaling

As stated previously, fixed-point processors have a much smaller dynamic range than their floating-point counterparts. It is due to this limitation that the Q15 representation of numbers is normally considered. For instance, a 16-bit multiplier can be used to multiply two Q15 numbers and produce a 32-bit product. Then the product can be stored in 32 bits or shifted back to 16 bits for storage or further processing.

When two Q15 numbers are being multiplied, which are in the range of −1 and 1, as discussed earlier, the product will be in the same range. However, when two Q15 numbers are added, the sum may fall outside this range, leading to an overflow. Overflows can cause major problems by generating erroneous results. When one is using a fixed-point processor, the range of numbers must be closely examined and, if necessary, be adjusted to compensate for overflows. The simplest correction method for avoiding overflows is scaling.

The idea of scaling is to scale down the system input before performing any processing and then to scale up the resulting output to the original size. Scaling can be applied to most filtering and transform operations. An easy way to achieve scaling is by shifting. Since a right shift of 1 is equivalent to a division by 2, we can scale the input repeatedly by 0.5 until all overflows disappear. The output can then be rescaled back to the total scaling amount.

As far as FIR and IIR filters are concerned, it is possible to scale coefficients to avoid overflows. Let us consider the output of an FIR filter $y[n] = \sum_{k=0}^{N-1} h[k]x[n-k]$, where h's denote coefficients or unit sample response terms and x's represent input samples. In the case of IIR filters, for a large enough N, the terms of the unit sample response become so small that they can be ignored. Let us suppose that x's are in Q15 format (i.e., $|x[n-k]| \leq 1$). Therefore, we can write $|y[n]| \leq \sum_{k=0}^{N-1} |h[k]|$.

This means that, to ensure no output overflow (i.e., $|y[n]| \leq 1$), the condition $\sum_{k=0}^{N-1} |h[k]| \leq 1$ must be satisfied. This condition can be satisfied by repeatedly scaling (dividing by 2) the coefficients or unit sample response terms.

5.5 Data Types in LabVIEW

The numeric data types in LabVIEW together with their symbols and ranges are listed in Table 5-1.

Table 5-1: Numeric Data Types in LabVIEW [4]

Terminal Symbol	Numeric Data Type	Bits of Storage on Disk
SGL	Single-precision, floating-point	32
DBL	Double-precision, floating-point	64
EXT	Extended-precision, floating-point	128
CSG	Complex single-precision, floating-point	64
CDB	Complex double-precision, floating-point	128
CXT	Complex extended-precision, floating-point	256
I8	Byte signed integer	8
I16	Word signed integer	16
I32	Long signed integer	32
U8	Byte unsigned integer	8
U16	Word unsigned integer	16
U32	Long unsigned integer	32
x	128-bit time stamp	<64.64>

Note that, other than the numeric data types shown in Table 5-1, there exist other data types in LabVIEW, such as cluster, waveform, and dynamic data type; see Table 5-2. For more details on all the LabVIEW data types, refer to [3,4].

Table 5-2: Other Data Types in LabVIEW [4]

Terminal Symbol	Data Type
<>	Enumerated type
TF	Boolean
abc	String
[ ]	Array—Encloses the data type of its elements in square brackets and takes the color of that data type.
cluster	Cluster—Encloses several data types. Cluster data types are brown if all elements in the cluster are numeric or pink if all the elements of the cluster are different types.
path	Path
dynamic	Dynamic—(Express VIs) Includes data associated with a signal and the attributes that provide information about the signal, such as the name of the signal or the date and time the data were acquired.
waveform	Waveform—Carries the data, start time, and dt of a waveform.
digital waveform	Digital waveform—Carries start time, delta x, the digital data, and any attributes of a digital waveform.
0101	Digital—Encloses data associated with digital signals.

(Continued)

Table 5-2: Other Data Types in LabVIEW [4]—Cont'd

Terminal Symbol	Data Type
	Reference number (refnum)
	Variant—Includes the control or indicator name, the data type information, and the data itself.
	I/O name—Passes resources you configure to I/O VIs to communicate with an instrument or a measurement device.
	Picture—Includes a set of drawing instructions for displaying pictures that can contain lines, circles, text, and other types of graphic shapes.

5.6 Bibliography

[1] J. Proakis and D. Manolakis, *Digital Signal Processing: Principles, Algorithms, and Applications*, Prentice-Hall, 1996.

[2] Texas Instruments, *TMS320C6000 CPU and Instruction Set Reference Guide*, Literature ID# SPRU189F, 2000.

[3] National Instruments, *LabVIEW Data Storage*, Application Note 154, Part Number 342012C-01, 2004.

[4] National Instruments, *LabVIEW User Manual*, Part Number 320999E-01, 2003.

Lab 5: Data Type and Scaling

Fixed-point implementation of a DSP system requires one to examine permissible ranges of numbers so that necessary adjustments are made to avoid overflows. The most widely used approach to cope with overflows is scaling. The scaling approach is covered in this lab.

L5.1 Handling Data Types in LabVIEW

In LabVIEW, the data type of exchanged data between two blocks is exhibited by the color of their connecting wires as well as their icons. A mismatched data type is represented by a coercion dot on a function or subVI input terminal, alerting that the input data type is being coerced into a different type. In general, a lower precision data value gets converted to a higher precision value. Coercion dots can lead to an increase in memory usage and run time [1]. Thus, it is recommended to resolve coercion dots in a VI.

An example exhibiting a mismatched data type is depicted in Figure L5-1. A double precision value and a 16-bit integer are wired to the input terminals of an Add

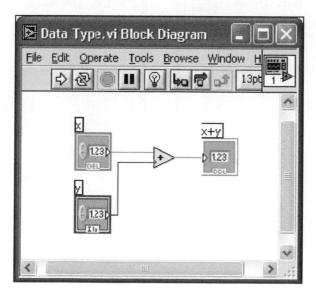

Figure L5-1: Data type mismatch.

function. As can be seen from this figure, a coercion dot appears at the y terminal of the Add function, since the input to this terminal is of 16-bit integer type, whereas the other input is of double-precision type.

Let us build the VI shown in Figure L5-1. Place an Add function and create two input controls by right-clicking and choosing **Create » Control** from the shortcut menu at each input terminal. By default, the data types of the two controls are set to double precision. In order to change the data type of the second Numeric Control, labeled y, right-click on the icon on the BD and select **Representation » Word**, which is represented by **I16**.

Create a Numeric Indicator by right-clicking on the x+y terminal of the Add function and choosing **Create » Indicator** from the shortcut menu. The data type of the newly created Numerical Indicator is double precision, since the addition of two double-precision values results in another double-precision value.

Let us switch to the FP of the VI to demonstrate the importance of specifying the correct data type to the Numeric Control/Indicator. If the value entered on the FP control does not match the data type specified by the Numeric Control/ Indicator, the input value is automatically converted into the data type specified by the Numeric Control. In the example shown in Figure L5-2, the

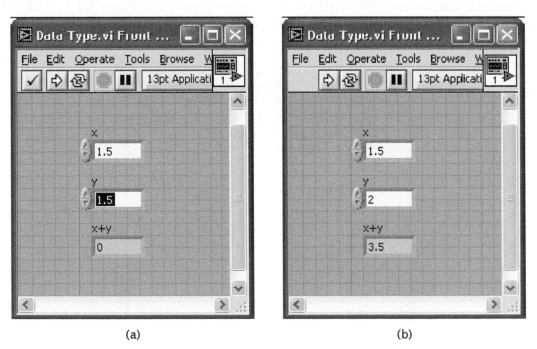

(a)   (b)

Figure L5-2: Data type conversion: (a) data typed in and (b) data are converted to 16-bit integer by LabVIEW.

value 1.5 is entered in both of the Numeric Controls on the FP. As can be seen, the entered value of the second Numeric Control, labeled y, automatically gets converted to a 16-bit integer or 2.

Coercion dots can be avoided if appropriate conversion functions are inserted in the BD. For example, as shown in Figure L5-3, the addition of a double precision and a 16-bit integer value is achieved without getting a coercion dot by inserting a To Double Precision Float function.

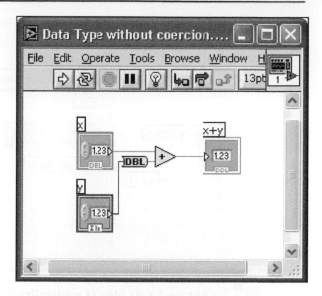

Figure L5-3: Data type conversion.

L5.2 Overflow Handling

An overflow occurs when the outcome of an operation is too large or too small for a processor to handle. In a 16-bit system, when one is manipulating integer numbers, they must remain in the range –32768 to 32767. Otherwise, any operation resulting in a number smaller than –32768 or larger than 32767 will cause overflow. For example, when 32767 is multiplied by 2, the result is 65534, which is beyond the representation limit of a 16-bit system.

Consider the BD shown in Figure L5-4. In this BD, samples of a sinusoidal signal having an amplitude of 30000 are multiplied by 2. To illustrate the overflow problem, the input values generated by the Sine Waveform VI are converted to word signed integers or 16-bit integers (I16). This is done by inserting a To Word Integer function (**Functions » Programming » Numeric » Conversion » To Word Integer**) at the output of the Get Waveform Component function. After the insertion of this function, the color of the wire connected to the output terminal of the function should appear blue, indicating integer data type.

In addition, the multiplicand constant should also be converted to the I16 data type to avoid a coercion dot. To achieve this, right-click on the Numeric Constant and select **Representation » Word**. As a result, the data type of the multiplication outcome or product is automatically set to word signed integer (I16).

It should be noted that the definition of data types is software/hardware dependent. For example, the "word" data type in LabVIEW has a length of 16 bits, whereas it

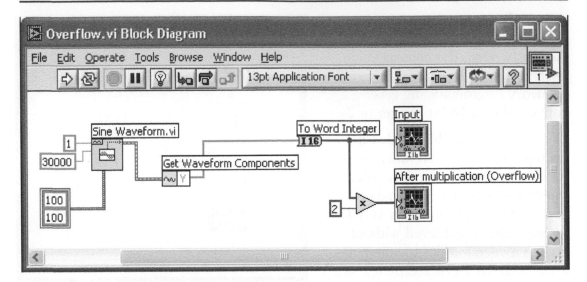

Figure L5-4: Signal multiplication data type conversion.

is 32 bits in the C6x DSP. That is, the "word" data type in LabVIEW is equivalent to the "short" data type in the C6x DSP [2].

The multiplication of a sinusoidal signal by 2 is expected to generate another sinusoidal signal with twice the amplitude. However, as seen from the FP in Figure L5-5, the result is distorted and clipped when the product is beyond the

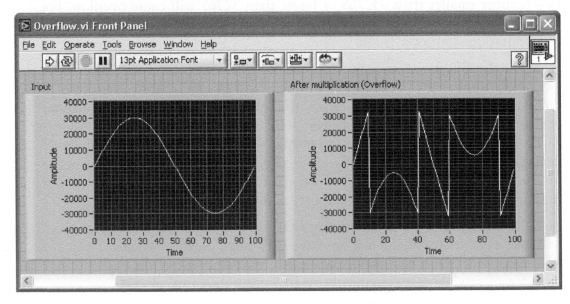

Figure L5-5: Signal distorted by overflow.

word integer (I16) range. Let us examine whether any overflow is caused by these multiplications. From Figure L5-5, one can see that the output signal includes wrong values due to overflows. For example, −5536 is shown to be the result of the multiplication of 30000 (the maximum input value) by 2, which is incorrect.

L5.2.1 Q-Format Conversion

Let us now consider the conversion of single- or double-precision values to Q-format. As shown in Figure L5-6, a double-precision input value is first checked to see whether it is in the range of −1 and 1. This is done by using the In Range and Coerce function (**Functions » Programming » Comparison » In Range and Coerce**). The input is scaled so that it falls within the range of 16-bit signed integer data type or −32768 to 32767, by multiplying it with its maximum allowable value 32768. Then, the product is converted to 16-bit signed integer data type by using a To Word Integer function. This ensures that the product falls within the range of the specified data type. In the worst case, the product gets clipped or saturated to the maximum or minimum allowable value.

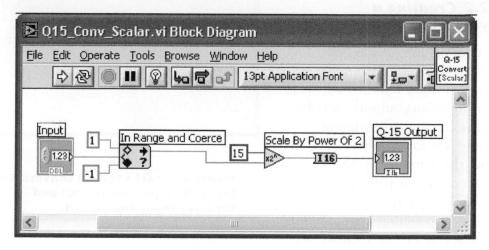

Figure L5-6: BD of Q15 format conversion.

Edit the icon of the VI as shown in Figure L5-6. The connector pane of the VI has one input and one output terminal. Assign the input terminal to the Numeric Control and the output terminal to the Numeric Indicator. Save the VI in a file named *Q15_Conv_Scalar.vi* and use it as a Q15 format converter for scalar inputs.

Next, the Q15_Conv_Scalar VI is modified to perform Q15 conversion for array type inputs/outputs as follows. As shown in Figure L5-7, one needs to replace the scalar numeric control and indicator with an array of controls and indicators, respectively. An array of controls or indicators can be created by first placing an Array shell (**Controls » Modern » Array, Matrix & Cluster » Array**) and then by dragging a control or indicator into it. The icon and the connector pane of the modified VI should be reconfigured accordingly. Save the modified VI in a file named *Q15_Conv_Array.vi*.

L5.2.2 Creating a Polymorphic VI

The two VIs just created are integrated into a polymorphic VI so that one VI can handle both scalar and array inputs/outputs. A polymorphic VI is a collection of multiple VIs for different instances having the same input and output connector pane [3]. The multiplication function is a good example of polymorphism, since it can be applied to two scalars, an array and a scalar, or two arrays.

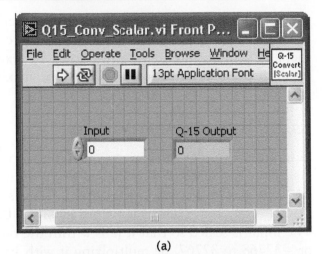

(a)

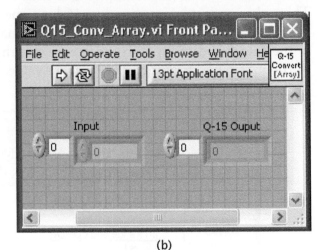

(b)

Figure L5-7: Q15 format conversion: (a) scalar input and output and (b) array input and output.

To create a polymorphic VI, select **File » New » VI » Polymorphic VI**. This brings up a Polymorphic VI window, as shown in Figure L5-8. Add the Q15_Conv_Scalar and Q15_Conv_Array VIs to include both the scalar and array cases. Edit the icon of the polymorphic VI as shown in Figure L5-8 and then save the VI as *Q15_Conv.vi*. This polymorphic VI is used for the remaining part of this lab.

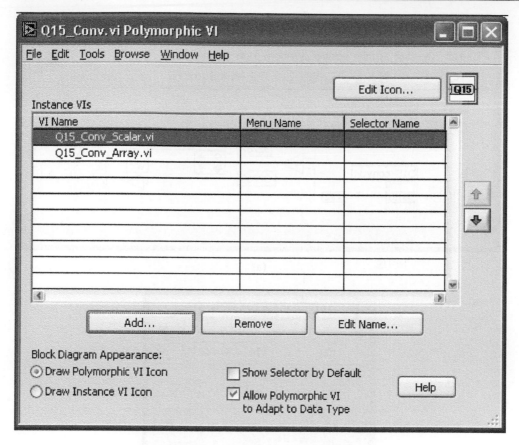

Figure L5-8: Creating a polymorphic VI.

Next, a VI is presented to show how the overflow is checked. In the BD shown in Figure L5-9(a), two input values in the range between −1 and 1 are converted to Q15 format by using the polymorphic Q15_Conv VI. These inputs are converted into a higher precision data type, e.g., long integer (I32), to avoid getting any saturation during their addition. Consider that LabVIEW automatically limits the output of numerical operations to the input range. The sum of the two input values is wired into the In Range and Coerce function to check whether they are in the allowable range. If the output does not fall in the range of I16, this indicates that an overflow has occurred. The FP in Figure L5-9(b) illustrates such an overflow.

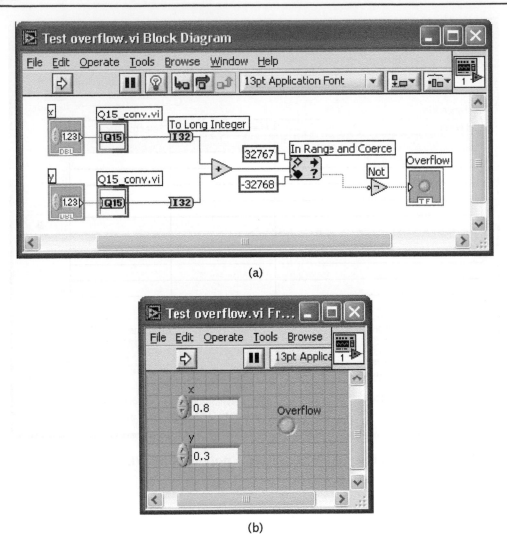

(a)

(b)

Figure L5-9: Test for overflow: (a) BD and (b) FP.

L5.3 Scaling Approach

Scaling is the most widely used approach to overcome the overflow problem. In order to see how scaling works, let us consider a simple multiply/accumulate operation. Suppose there are four constants or coefficients that need to be multiplied with samples of an input signal. The worst possible overflow case would be the one in which all the multiplicands C_k's and $x[n]$'s are 1's. For this case, the result $y[n]$ will be 4, given that $y[n] = \sum_{k=1}^{4} C_k x[n-k]$. Assuming that we have control only over the input signal, the input samples should be scaled for the result or sum $y[n]$ to fall in the allowable range. A single half-scaling, or division by 2, reduces the input samples by one-half, and a double half-scaling reduces them further by one-quarter. Of course, this leads to less precision, but it is better than getting an erroneous outcome.

A simple method to implement the scaling approach is to create a VI that returns the necessary amount of scaling on the input. For multiply/accumulate types of operations, such as filtering or transforms, the worst case is multiplications and additions of all 1's. This means that the required number of scaling is dependent on the number of additions in the summation. To examine the worst case, one needs to obtain the required number of scaling so that all overflows disappear. This can be achieved by building a VI to compute the required number of scaling. For the example covered in this lab, such a VI is shown in Figure L5-10 and is implemented in a hybrid fashion utilizing the MathScript node feature of LabVIEW.

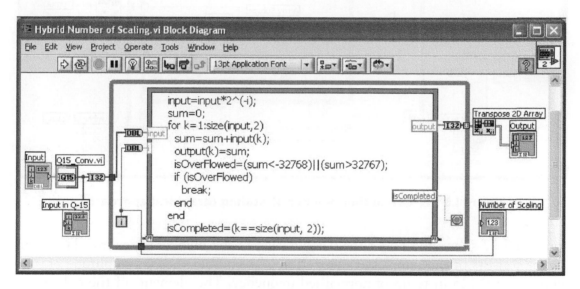

Figure L5-10: Computing number of scaling (hybrid approach).

Here, the input is first converted into Q15 format via the Q15_Conv VI. Inside the outer loop (LabVIEW While Loop), the input is scaled. In each iteration of the inner loop (MathScript For Loop), a new input sample is taken into consideration and a summation is obtained. Then, the summation value is compared with the minimum and maximum values in the allowable range. If the summation value does not fall into this range, the inner loop is stopped, and the input samples are scaled down for a next iteration. The number of scaling is also counted. After the input is scaled, the summation is repeated. If another overflow occurs, the input sample is scaled down further. This process is continued until no overflow occurs. The final number of scaling is then displayed. Care must be taken not to scale the input too many times; otherwise, the input signal gets buried in the quantization noise. Here, the auto-indexing is enabled to collect the output samples into an array corresponding to each `While Loop`. As a result, a 2D array is generated.

Figure L5-11 shows the same method to compute the required number of scaling without using the MathScript node feature.

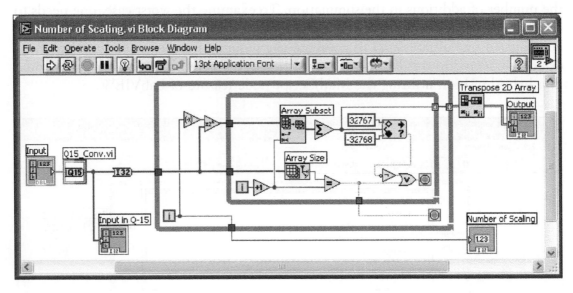

Figure L5-11: Computing number of scaling (graphical approach).

Figure L5-12 shows the FP for the computation of the number of scaling. The input signal consists of the samples of one period of a sinusoid with an amplitude of 0.8 sampled at 0.125 in terms of normalized frequency. The elements of the column shown in the output indicator represent the accumulated sums of the input samples. Notice that an overflow occurs at the third summation, since the value is greater than the maximum value of a 16-bit signed integer, i.e., 32767. The overflow disappears if the input is scaled down once by one-half. Thus, in this example, the required number of scaling to avoid any overflow is one.

It is worth mentioning that, in addition to scaling the input, one also can scale the filter coefficients or constants in convolution type of operations so that the outcome is forced to stay within the allowable range. In this case, the worst case for the input samples is assumed to be one. Note that scaling down the coefficients

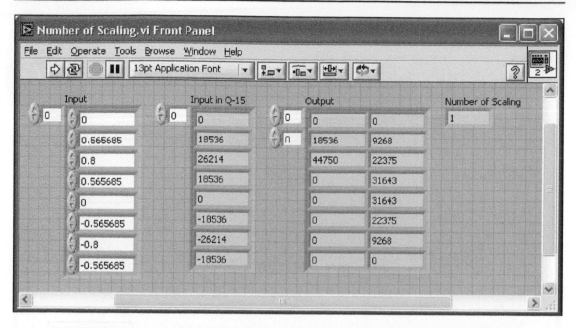

Figure L5-12: Number of scaling for one period of sinusoidal signal.

by one-half is equivalent to scaling down the input samples by one-half. An example of fixed-point digital filtering as well as coefficient scaling is examined in the following section.

L5.4 Digital Filtering in Fixed-Point Format

The analysis of overflow and scaling discussed in the preceding sections is repeated here by using the DFD Fixed-Point Tools VIs. These VIs allow the quantization of filter coefficients and the fixed-point simulation of digital filters. As an example of fixed-point digital filtering, the FIR lowpass filter designed in Lab 4 is revisited here.

L5.4.1 Design and Analysis of Fixed-Point Digital Filtering System

To design an FIR filter, place the DFD Classical Filter Design Express VI on the BD and enter the filter specifications on the configuration dialog box of this Express VI. Four VIs are used as part of the DFD toolkit to display the filter response, as shown in Figure L5-13.

Let us examine each object of this BD. The DFD FXP Quantize Coef VI **(Functions » Addons » Digital Filter Design » Fixed-Point Tools » DFD FXP Quantize**

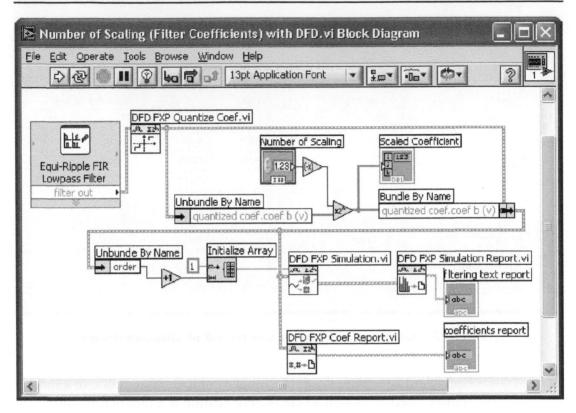

Figure L5-13: Computing number of scaling with DFD toolkit.

Coef) quantizes the filter coefficients according to the specified options. By default, a 16-bit word length is used for quantization. The b_k coefficients of the filter are unbundled by using the Unbundle By Name function. Right-click on the Unbundle By Name function and choose **Select Item » quantized coef » coef b(v)** from the shortcut menu. Next, the quantized coefficients are scaled down to match the number of scaling specified in the Numeric Control. The use of an array of indicators labeled as Scaled Coefficient is optional. This can be used to easily export the filter coefficients to other VIs. The original filter coefficients are replaced with the scaled coefficients by using the Bundle by Name function.

The DFD FXP Simulation VI (**Functions » Addons » Digital Filter Design » Fixed-Point Tools » DFD FXP Simulation**) simulates the filter operation and generates its statistics using the fixed-point filter coefficient set. The filter statistics include min and max value, number of overflow/underflow, and number of operation. A text report of the filter statistics is generated via the DFD FXP Simulation Report VI (**Functions » Addons » Digital Filter Design » Fixed-Point Tools » DFD FXP Simulation Report**) and displayed in the String Indicator. The DFD FXP Coef Report VI (**Functions » Addons » Digital Filter Design » Fixed-Point Tools » DFD FXP Coef Report**) generates a text report on the quantized filter coefficients.

In order to consider the worst case scenario, one can create an array of all ones and wire them to act as the input of the filter. The length of this array is determined by the number of filter coefficients. The simulated result is shown in Figure L5-14.

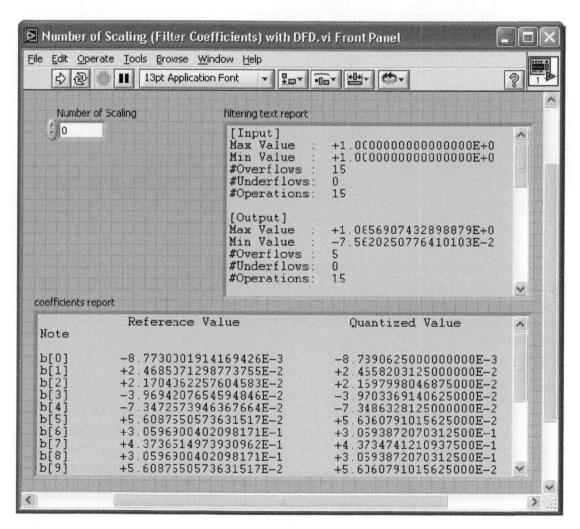

Figure L5-14: Fixed-point analysis using DFD toolkit (no scaling).

Table L5-1: Scaling Example

C_k	$\sum C_k$	$\frac{C_k}{2}$	$\sum \frac{C_k}{2}$
−0.00878906	−0.00878906	−0.00439453	−0.00439453
0.0246582	0.01586914	0.0123291	0.00793457
0.021698	0.03756714	0.010849	0.01878357
−0.03970337	−0.00213623	−0.01985168	−0.00106811
−0.07348633	−0.07562256	−0.03674316	−0.03781127
0.05606079	−0.01956177	0.0280304	−0.00978087
0.30593872	0.28637695	0.15296936	0.14318849
0.43734741	0.72372436	0.21867371	0.3618622
0.30593872	**1.02966308**	0.15296936	0.51483156
0.05606079	**1.08572387**	0.0280304	0.54286196
−0.07348633	**1.01223754**	−0.03674316	0.5061188
−0.03970337	0.97253417	−0.01985168	0.48626712
0.021698	0.99423217	0.010849	0.49711612
0.0246582	**1.01889037**	0.0123291	0.50944522
−0.00878906	**1.01010131**	−0.00439453	0.50505069

As one can see in this figure, five overflows are reported at the output for the no scaling case. The sums of the filter coefficients C_k's are listed in Table L5-1 with the overflows in bold.

Now, enter 1 as the number of scaling and run the VI. The outcome of this simulation after scaling by one-half is shown in Figure L5-15. No overflow is observed after this scaling. The scaled coefficient set is also shown in this figure. In addition, the sums of the scaled coefficients are listed in Table L5-1 indicating no overflow.

L5.4.2 Filtering System

As stated previously, the coefficients of the FIR filter need to be scaled by one-half to avoid overflows. For simplicity, two arrays of constants containing the scaled filter coefficients are used on the BD, as shown in Figure L5-16. One way to create an array of constants corresponding to the filter coefficients is to change an array of indicators to an array of constants. To do this, one can copy an icon of an indicator, labeled as Scaled Coefficient in Figure L5-13, is copied to the BD

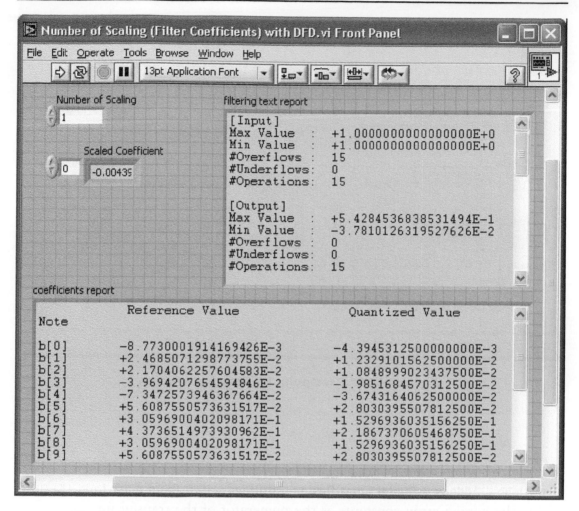

Figure L5-15: Fixed-point analysis using DFD toolkit (one scaling).

of a new VI. Make sure that the array of indicators displays the coefficients before it is copied to a new VI. Right-click on the icon of the indicator on the BD and choose **Change to Constant**. This way, an array of constants containing the filter coefficients is created.

A filter object is created from the coefficient or transfer function of the filter. This is done by placing the DFD Build Filter from TF VI

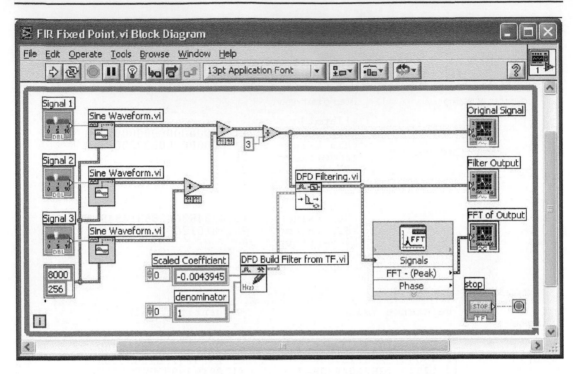

Figure L5-16: Fixed-point FIR filtering system.

(**Functions » Addons » Digital Filter Design » Utilities » DFD Build Filter from TF**) and wiring the copied array constants as the numerator of the transfer function. Note that the denominator of the FIR transfer function is a single element array of size 1. Once the filter cluster is created, it is wired to the DFD Filtering VI to carry out the filtering operation.

The input to the DFD Filtering VI consists of three sinusoidal signals. The summed input is divided by 3 to make the range of the input values between –1 and 1 and then is connected to the FIR filter. The input and output signals are shown in Figure L5-17. It can be observed that the fixed-point version of the filter operates exactly the same way as the floating-point version covered in Lab 4. The difference in the scales is due to the use of the one-half scaled filter coefficients and one-third scaled input values.

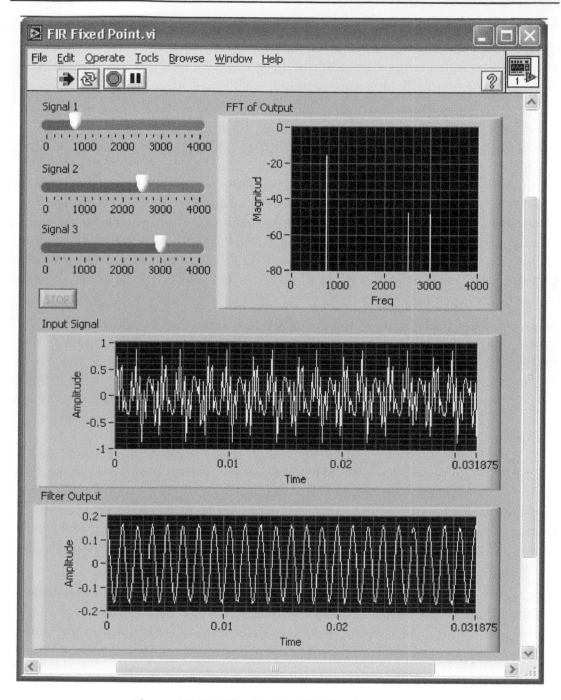

Figure L5-17: Fixed-point FIR filtering output.

L5.4.3 Fixed-Point IIR Filter Example

Considering that the stability of an IIR filter is sensitive to the precision used, an example is provided here to demonstrate this point. This example involves the fixed-point versions of an IIR filter corresponding to different filter forms.

Let us consider an IIR lowpass filter with the following specifications: passband response $= 0.1$ dB, passband frequency $= 1200$ Hz, stopband attenuation $= 30$ dB, stopband frequency $= 2200$ Hz, and sampling rate $= 8000$ Hz. The default form of the IIR filter designed by the DFD Classical Filter Design Express VI is the second-order cascade form. The filter can be converted to the direct form by using the DFD Convert Structure VI (**Functions » Addons » Digital Filter Design » Conversion » DFD Convert Structure**). The DFD Convert Structure VI provides a total of 23 forms as the target structure. A Ring Constant is created by right-clicking on the target structure terminal of the VI and then choosing **Create » Constant**. Click on the created Ring Constant and select IIR Direct Form II as the target structure.

Next, the filter coefficients in the direct form are quantized by using the DFD FXP Quantize Coef VI. Notice that 16 bits is the default word length of the fixed-point representation in the absence of a configuration cluster constant. Different configurations of quantization can be set by creating and wiring a cluster constant at the Coefficient quantizer terminal of the DFD FXP Quantize Coef VI. The floating-point (unquantized) and fixed-point (quantized) filter clusters are wired to the DFD Filter Analysis Express VIs, which are configured to create the magnitude responses. These magnitude responses are placed into one waveform graph. This is done by creating a graphical indicator at the magnitude terminal of one of the Express VIs and then by wiring the output of the other Express VI to the same waveform graph. A Merge Signal function gets automatically located on the BD. This normally occurs when two or more dynamic data type wires are merged. The BD of the fixed-point IIR filter is shown in Figure L5-18.

The quantized filter object is also wired to a DFD FXP Coef Report VI (**Functions » Addons » Digital Filter Design » Fixed-Point Tools » DFD FXP Coef Report**) to generate a text report. This report provides reference coefficients, quantized coefficients, and note sections such as overflow/underflow.

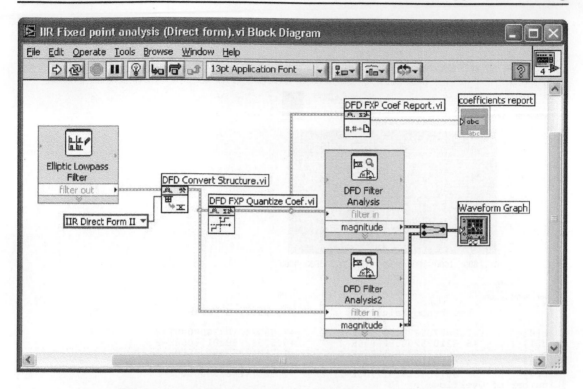

Figure L5-18: BD of fixed-point IIR filter in direct form.

The FP of the VI after running the fixed-point filter is shown in Figure L5-19. Notice that the line style of the fixed-point plot is chosen as dotted for comparison purposes. This is done by right-clicking on the label or plotting in the plot legend and choosing **Line Style** from the shortcut menu.

From Figure L5-19, the magnitude response of the fixed-point version of the IIR filter is seen to be quite different than its floating-point version. This is due to the fact that one underflow and one overflow occur in the filter coefficients, causing the discrepancies in the responses.

Next, let us examine the fixed-point version of the IIR filter in the second-order cascade form. This can be achieved simply by removing the DFD Convert

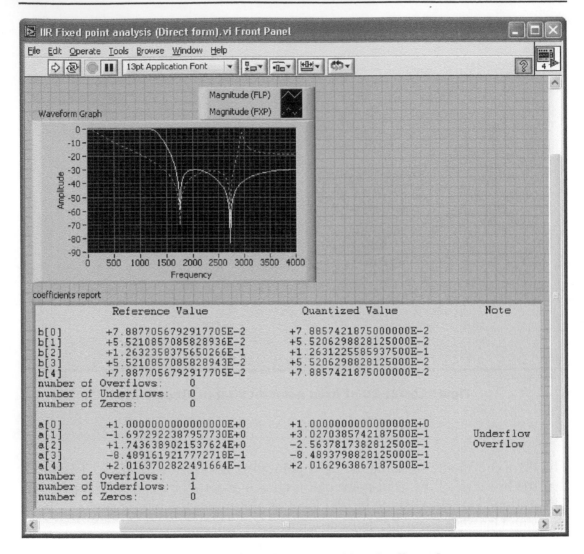

Figure L5-19: FP of fixed-point IIR filter in direct form.

`Structure` VI from the BD shown in Figure L5-18. The magnitude responses of the floating-point and fixed-point versions are illustrated in Figure L5-20. These magnitude responses appear to be identical. Also, no overflow or underflow is observed. This indicates that the effect of quantization can be minimized by using the second-order cascade form.

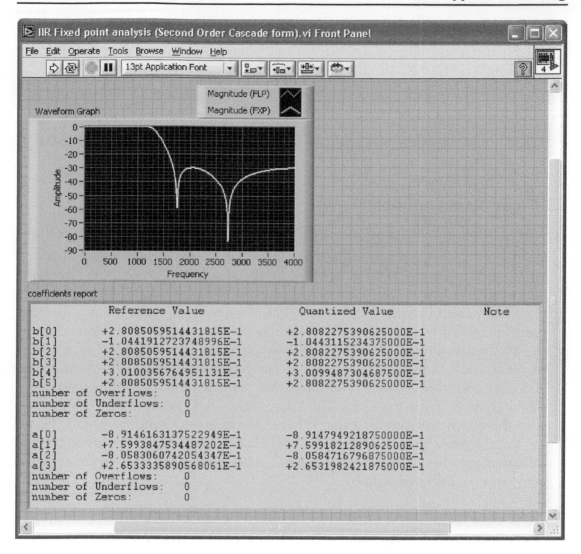

Figure L5-20: FP of fixed-point IIR filter in cascade form.

L5.5 Bibliography

[1] National Instruments, *LabVIEW Data Storage, Application Note 154*, Part Number 342012C-01, 2004.

[2] Texas Instruments, *TMS320C6000 CPU and Instruction Set Reference Guide*, Literature Number: SPRU189F, 2000.

[3] National Instruments, *LabVIEW User Manual*, Part Number 320999E-01, 2003.

L5.6 Lab Experiments

Perform the following experiments with and without using the MathScript feature.

1. Build a VI to design an IIR elliptic bandpass floating-point filter based on a sampling frequency of 8000 Hz. Set the passband response as 0.1dB and stopband attenuation as 60 dB. Use a passband frequency range of 1600–2400 Hz and a stopband frequency range of 1200–2800 Hz. Generate the fixed-point filter using the DFD toolkit by quantizing the coefficients of the floating-point filter. Convert the fixed-point filter coefficients to Q15 format and find out the number of scaling required to avoid overflow. Compare the magnitude response of the filter before and after coefficient scaling. Use the scaled coefficients to filter the composite signal given by Equation (5.9). This composite signal consists of three sinusoids with the number of samples being 128. Set the frequency range as 0–3500 Hz for each of the frequencies f_1 Hz, f_2 Hz, and f_3 Hz. Generate the composite signal and display it together with the filtered signal. Also, display the FFT spectrum of the composite and filtered signals.

$$x(t) = \left(\frac{\sin(2\pi f_1 t) + \cos(2\pi f_2 t) + \sin(2\pi f_3 t)}{5} \right) \qquad (5.9)$$

2. Build a VI to design an IIR Chebyshev bandpass floating-point filter based on a sampling frequency of 8000 Hz. Set the passband response as 0.1dB and stopband attenuation as 60 dB. Use a passband frequency range of 1300–2700 Hz and a stopband frequency range of 1000–3000 Hz. Generate the fixed-point filter using the DFD toolkit by quantizing the coefficients of the floating-point filter. Convert the fixed-point filter coefficients to Q15 format and find out the number of scaling required to avoid overflow. Compare the magnitude response of the filter before and after coefficient scaling. Use the scaled coefficients to filter the composite signal given by Equation (5.10). This composite signal consists of three sinusoids

with the number of samples being 128. Set the frequency range as 0–4000 Hz for each of the frequencies f_1 Hz, f_2 Hz, and f_3 Hz. Generate the composite signal and display it together with the filtered signal. Also, display the FFT spectrum of the composite and filtered signals.

$$x(t) = \left(\frac{\cos\,(2\pi f_1 t) + \sin\,(2\pi f_2 t) + \cos\,(2\pi f_3 t)}{6} \right) \qquad (5.10)$$

3. Build a VI to compute the equivalent decimal magnitude of 16-bit integers using the following formats: (i) Q15, (ii) Q12, and (iii) Q10. For example, in the case of Q12 format, the 4 MSB bits of a 16-bit integer should correspond to the integer part of the number and the remaining 12 bits to the fractional part of the number. For negative integers, first generate the 2's-complement bits; then use these bits to compute the equivalent decimal magnitude, followed by negation to obtain the final result.

4. Build a VI to compute the equivalent decimal magnitude of 32-bit integers using the following formats: (i) Q25, (ii) Q23, and (iii) Q20. For example, in the case of Q25 format, the 7 MSB bits of a 32-bit integer should correspond to the integer part of the number and the remaining 25 bits to the fractional part of the number. For negative integers, first generate the 2's-complement bits; then use these bits to compute the equivalent decimal magnitude, followed by negation to obtain the final result.

6

Adaptive Filtering

Adaptive filtering is used in many applications including noise cancellation and system identification. In most cases, the coefficients of an FIR filter are modified according to an error signal in order to adapt to a desired signal. In this chapter, a system identification and a noise cancellation system are presented wherein an adaptive FIR filter is used.

6.1 System Identification

In system identification, the behavior of an unknown system is modeled by accessing its input and output. An adaptive FIR filter can be used to adapt to the output of the unknown system based on the same input. As indicated in Figure 6-1, the difference in the output of the system, $d[n]$, and the output of the adaptive FIR filter, $y[n]$, constitutes the error term, $e[n]$, which is used to update the coefficients of the filter.

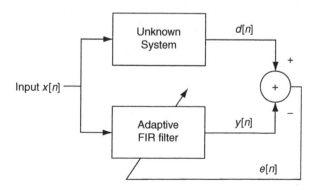

Figure 6-1: System identification system.

The error term, or the difference between the outputs of the two systems, is used to update each coefficient of the FIR filter according to the equation (known as the least mean square, or LMS, algorithm [1])

$$h_n[k] = h_{n-1}[k] + \delta e[n]x[n-k] \tag{6.1}$$

where h's denote the unit sample response or FIR filter coefficients, and δ denotes a step size. This adaptation causes the output $y[n]$ to approach $d[n]$. A small step size will ensure convergence but result in a slow adaptation rate. A large step size, though faster, may lead to skipping over the solution.

6.2 Noise Cancellation

A system for adaptive noise cancellation has two inputs consisting of a noise-corrupted signal and a noise source. Figure 6-2 illustrates an adaptive noise cancellation system. A desired signal $s[n]$ is corrupted by a noise signal $v_1[n]$, which originates from a noise source signal $v_0[n]$. Bear in mind that the original noise source signal gets altered as it passes through an environment or channel whose characteristics are unknown. For example, this alteration can be in the form of a lowpass filtering process. Consequently, the original noise signal $v_0[n]$ cannot be simply subtracted from the noise-corrupted signal, as there exists an unknown dependency between the two noise signals, $v_1[n]$ and $v_0[n]$. The adaptive filter is thus used to provide an estimate for the noise signal $v_1[n]$.

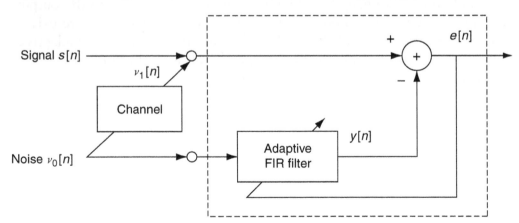

Figure 6-2: Noise cancellation system.

The weights of the filter are adjusted in the same manner stated previously. The error term of this system is given by

$$e[n] = s[n] + v_1[n] - y[n] \tag{6.2}$$

The error $e[n]$ approaches the signal $s[n]$ as the filter output adapts to the noise component of the input $v_1[n]$. To obtain an effective noise cancellation system, one should place the sensor for capturing the noise source adequately far from the signal source.

6.3 Bibliography

[1] S. Haykin, *Adaptive Filter Theory*, Prentice-Hall, 1996.

Lab 6: Adaptive Filtering Systems

This lab covers adaptive filtering by presenting two adaptive systems using the LMS algorithm, one involving system identification and the other noise cancellation.

L6.1 System Identification

A seventh-order IIR bandpass filter having a passband from $\pi/3$ to $2\pi/3$ radians is used here to act as the unknown system. An adaptive FIR filter is designed to adapt to the response of this system.

L6.1.1 Least Mean Square (LMS) Algorithm

Figure L6-1 shows the BD of the LMS VI, which is built by using the MathScript Node feature. The inputs of this VI include desired output (Input 1), array of samples in a previous iteration (x[n]), input to the unknown system (Input 2), step size, and filter coefficient set of the previous iteration (h[n]) ordered from top to bottom. The outputs of this VI include updated array (x[n + 1]), error term, FIR filter output, and updated filter coefficient set (h[n + 1]) ordered from top to bottom.

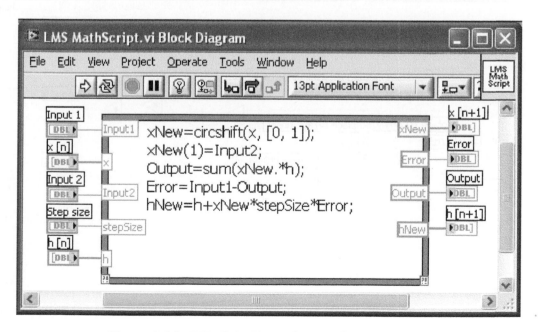

Figure L6-1: BD of LMS VI using MathScript Node.

161

The LMS algorithm inside the MathScript Node consists of five lines of code. Lines 1 and 2 create a circular buffer to read one sample at a time. Line 3 computes the filter output using a dot product. Line 4 generates the error signal, and line 5 updates the filter coefficients as stated in Equation (6.1).

At this point, it is worth mentioning that it is possible to use the same code within a MATLAB Script Node (**Functions » Mathematics » Script & Formula » Script Nodes » MATLAB Script Node**), with the difference that this approach requires the installation of MATLAB. In MathScripting, M-files get interpreted in LabVIEW, whereas in MATLAB scripting, they are interpreted in the MATLAB environment.

As an alternative to scripting or the textual approach, the LMS VI can be built graphically, as shown in Figure L6-2. The two array functions, Replace Array Subset (**Functions » Programming » Array » Replace Array Subset**) and Rotate 1D Array (**Functions » Programming » Array » Rotate 1D Array**), act as a circular buffer where the input sample at index 0 gets replaced by a new incoming sample. To perform point-by-point processing, the FIR Filter PtByPt VI (**Functions » Signal**

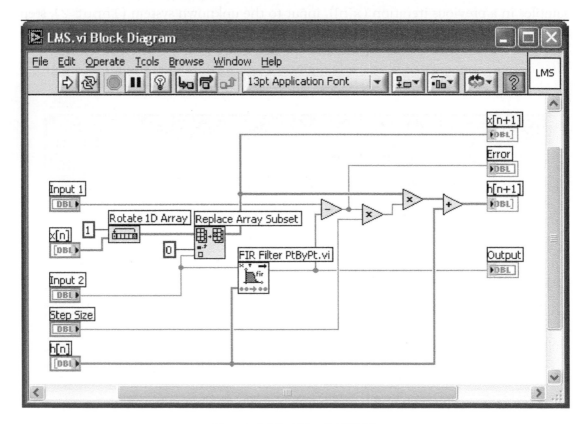

Figure L6-2; BD of LMS VI.

Processing » Point By Point » Filters PtByPt » FIR Filter PtByPt) is used. This VI requires a single-element input and a coefficient array.

The `Subtract` function on the BD calculates the error or the difference between the desired signal and the output of the adaptive FIR filter. This error is multiplied by the step size δ and then by the elements in the input buffer to obtain the coefficient updates. Next, these updates are added to the previous coefficients' $h[n]$'s to compute the updated coefficients' $h[n+1]$'s as stated in Equation (6.1).

The icon of the LMS VI is edited as shown in Figure L6-2. The connector pane of the VI is also modified as shown in Figure L6-3 for it to be used as a subVI.

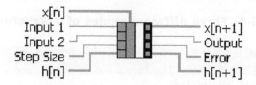

Figure L6-3: Connector pane of LMS VI.

L6.1.2 Waveform Chart

A waveform graph plots an array of samples, whereas a waveform chart takes one or more samples as its input and maintains a history so that a trajectory can be displayed, similar to an oscilloscope.

There are three different updating modes in a waveform chart. They include a Strip chart, Scope chart, and Sweep chart. The Strip chart mode provides a continuous data display. Once the plot reaches the right border, the old data are scrolled to the left, and new data are plotted on the right edge of the plotting area. The Scope chart mode provides a data display from left to right. Then, it clears the plot and resumes it from the left border of the plotting area. This is similar to data display on an oscilloscope. The Sweep chart mode functions similar to the Scope chart mode except the old data are not cleared. The old and new data are separated by a vertical line. These three modes are illustrated in Figure L6-4. They can be configured by right-clicking on the plot area and then by selecting **Advanced » Update Mode**.

The length of data displayed on the chart is changeable. To change it, right-click on the plot area and select **Chart History Length**. This brings up a dialog box to enter data length.

L6.1.3 Shift Register and Feedback Node

The BD of the overall adaptive system is shown in Figure L6-5. The figure shows two Feedback Nodes denoted by ▣. A Feedback Node is used to transfer data

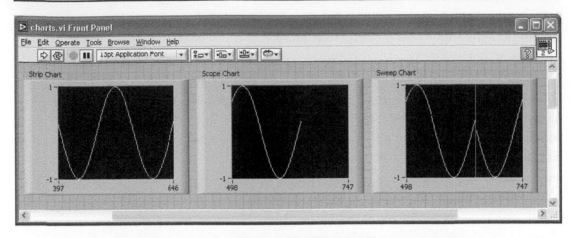

Figure L6-4: Three different modes of a waveform chart.

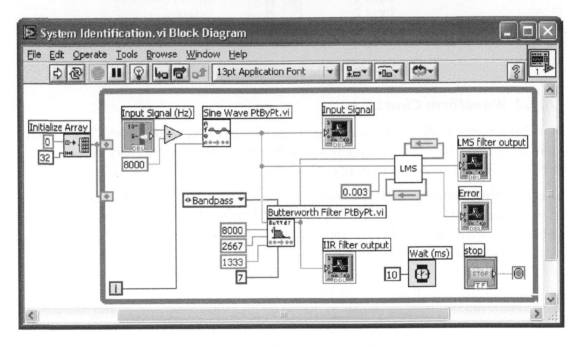

Figure L6-5: BD of system identification.

from one iteration to a next iteration inside a For Loop or a While Loop. A Feedback Node gets automatically generated when the output of a VI is wired to its input inside a loop structure. By default, an initializer terminal is added onto the left border of the loop for each Feedback Node. An initializer terminal is used to initialize the value to be transferred by the Feedback Node. If no initialization is needed, the terminal can be removed by right-clicking on it and unchecking **Initializer terminal.**

A Feedback Node can be replaced by a Shift Register. To achieve this, right-click on the Feedback Node. Then choose **Replace with Shift Register**. This adds a Shift Register at both sides of the `While Loop`. Also, the Shift Registers get wired to the terminals of the `LMS` subVI.

In the BD shown in Figure L6-5, a sinusoidal wave is generated to serve as the input signal, and a `Butterworth Filter PtByPt` VI (**Functions » Point By Point » Filters PtByPt » Butterworth Filter PtByPt**) is used to act as the unknown system. The filter coefficient array and the input data array are passed from one iteration to a next iteration by the Feedback Nodes to update the filter coefficients via the LMS algorithm. Both of these arrays are initialized with 32 zero values, considering that the number of filter taps is 32. This is done by wiring an `Initialize Array` function (**Functions » Programming » Array » Initialize Array**) to the initializer terminal. The initialization array is configured to be of length 32 containing zero values.

For the step size of the LMS algorithm, a `Numeric Constant` is created and wired. This value can be adjusted to control the speed of adaptation. In this example, 0.003 is used. Also, a `Wait (ms)` function is placed in the `While Loop` to delay the execution of the loop.

In addition to the programming approaches mentioned previously, a C DLL can be used to perform the function of the LMS VI. The following C source code can be used for this purpose.

```
#include <windows.h>
#include <string.h>
#include <ctype.h>

BOOL WINAPI DllMain (
    HANDLE  hModule,
    DWORD dwFunction,
    LPVOID lpNot)
{
    return TRUE;
}
/* This function implements LMS algorithm */
_declspec (dllexport) double LMS(double input1, double input2,
double stepSize, double *x, double *h, double *error)
{
    int i;
    int bufferLength=32;
    double output=0;
```

```
    // shift data in the input buffer
    for (i=bufferLength-1;i>0; i--)
    {
        x[i]=x[i-1];
    }
    x[0]=input2;
    // calculate output
    output=0;
    for (i=0; i<bufferLength; i++)
    {
        output=output+x[i]*h[i];
    }
    // calculate error
    error[0]=input1-output;
    // update coefficients
    for (i=0; i<bufferLength; i++)
    {
        h[i]=h[i]+stepSize*error[0]*x[i];
    }
 return output;
}
```

To incorporate the C DLL in the BD shown in Figure L6-6, one needs to use a Call Library Function Node to call it. Notice that two sets of Initialize Array VI are used: one for the input buffer (x[n]) and one for the filter coefficients

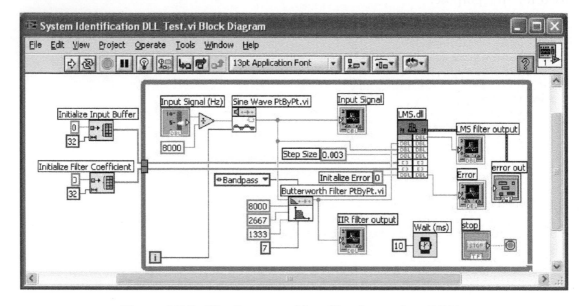

Figure L6-6: BD of system identification using C DLL.

(h[n]). It is necessary to separate the two initializations for this case, unlike the array initialization shown in Figure L6-5, because these two arrays are passed by pointers. Having the same pointer causes computation error. The input variable error is also passed by a pointer because it is desired to see this variable as an output.

As shown in Figure L6-7, the output of the adaptive LMS filter adapts to the output of the Butterworth IIR filter and thus the error between the outputs diminishes.

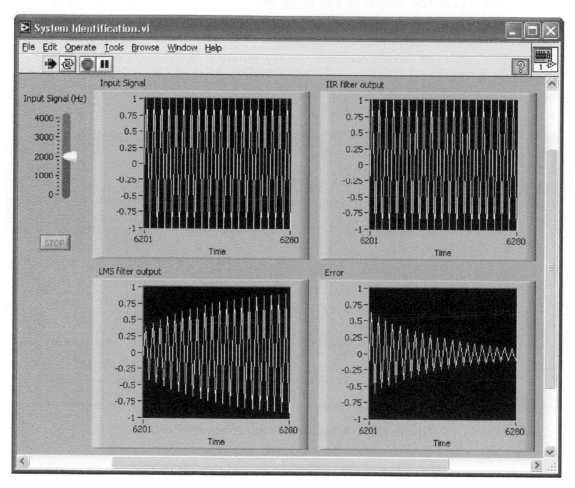

Figure L6-7: FP of system identification.

L6.2 Noise Cancellation

The design of a noise cancellation system is achieved similar to the system identification system mentioned previously. A noise cancellation system takes two inputs: a noise corrupted input signal and a reference noise signal. The BD of the adaptive noise cancellation system is shown in Figure L6-8. Again, as in the system identification example, the point-by-point processing feature is employed here. This requires using a Get Waveform Components function together with an Index Array function at the output of the noise and signal sources. The number of samples of the waveforms generated by the three Sine Waveform VIs is set to 1 for performing point-by-point processing. The data type of the Y component is still of array type with size 1. The Index Array function is used to extract a scalar element from the array. This ensures that the numerical operations are done in a point-by-point fashion.

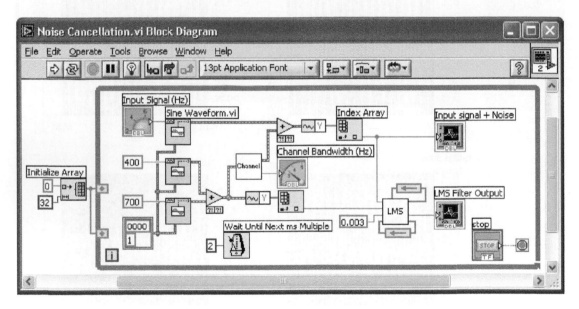

Figure L6-8: BD of noise cancellation system.

To be able to observe the adaptability of the system, one can add a time-varying channel. The noise source, which consists of the sum of two sinusoidal waveforms (400 and 700 Hz), is passed through the channel before it is added to the input signal. The BD of the time-varying channel is shown in Figure L6-9.

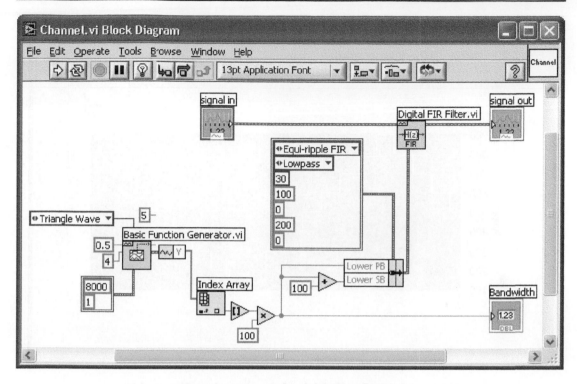

Figure L6-9: Time-varying channel.

The channel consists of an FIR lowpass filter with its passband and stopband varying according to a discretized triangular waveform. The reason for the discretization is to give the LMS algorithm enough time to adapt to the noise signal. The characteristic of the channel is varied with time by swinging the filter passband from 100 to 900 Hz. The bandwidth of the time-varying Channel VI is illustrated in Figure L6-10.

The waveform graph shown in Figure L6-5 indicates that the adaptive filter adapts to its input by cancelling out the noise component as the characteristic of the channel is changed. As illustrated in Figure L6-5, the input signal to the system is a 50 Hz sinusoid, and the noise varies in the range of 100–900 Hz. The step size δ may need to be modified depending on how fast the system is converging. It is necessary to make sure that the characteristic of the channel is not changing so fast, giving the adaptive filter adequate time to adapt to

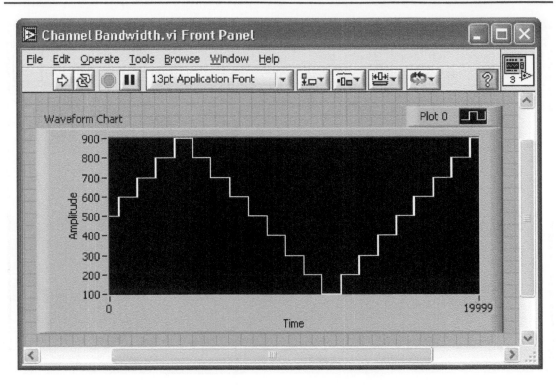

Figure L6-10: Bandwidth of time-varying Channel VI.

the noise signal passed through it. As one can see in Figure L6-11, the system adapts to the noise signal before the channel changes.

Note that the noise-cancelled signal is available from the Error terminal of the LMS VI. If a DC input signal, i.e., a 0 Hz signal, is applied to the system, the output of the adaptive filter becomes the error between the noise signal passed through the channel and the reference noise signal. This is illustrated in Figure L6-12.

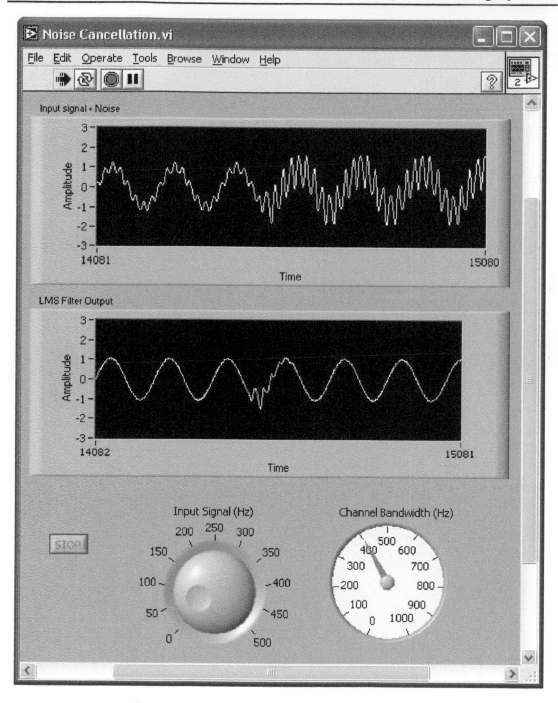

Figure L6-11: FP of noise cancellation system.

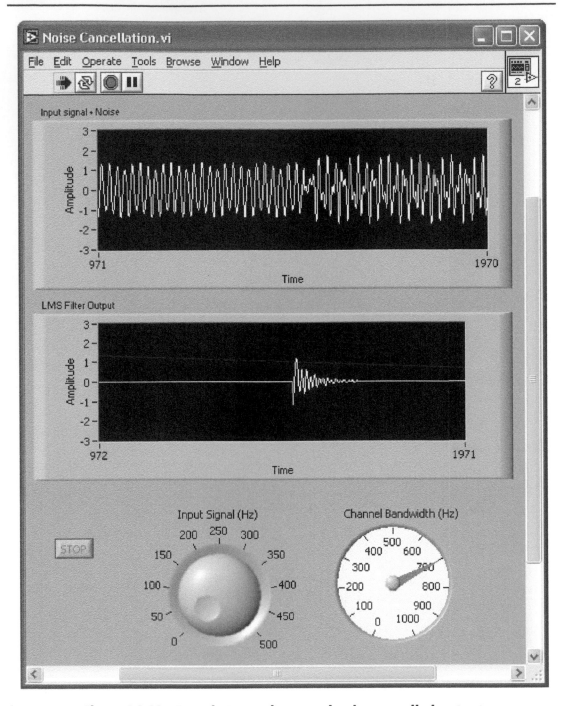

Figure L6-12: Error between input and noise-cancelled output.

L6.3 Lab Experiments

1. Build a VI graphically to implement the inverse system identification problem shown in Figure L6-13 by modifying the system identification VI appearing in Figure L6-5. Generate the desired signal by setting the delay equal to one-half the order of the unknown system. Verify the inverse system identification VI for the system orders 12 and 16.

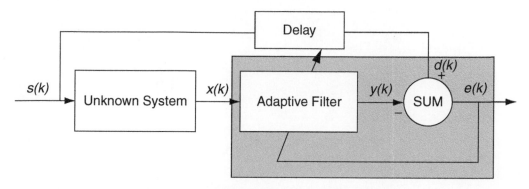

Figure L6-13: Inverse system identification.

2. Build a VI to implement the LMS VI shown in Figure L6-2 in a hybrid fashion by using the MathScript feature.

3. Build a VI to implement the time varying channel shown in Figure L6-9 by using the MathScript feature.

4. Build a VI to implement and verify the noise cancellation system shown in Figure L6-8 in a hybrid fashion as in (2) and (3).

Frequency Domain Processing

Transformation of signals to the frequency domain is widely used in signal processing. In many cases, such transformations provide a more effective representation and a more computationally efficient processing of signals as compared to time domain processing. For example, due to the equivalency of convolution operation in the time domain to multiplication in the frequency domain, one can find the output of a linear system by simply multiplying the Fourier transform of the input signal by the system transfer function.

This chapter presents an overview of three widely used frequency domain transformations, namely fast Fourier transform (FFT), short-time Fourier transform (STFT), and discrete wavelet transform (DWT). More theoretical details regarding these transformations can be found in many signal processing textbooks, e.g. [1].

7.1 Discrete Fourier Transform (DFT) and Fast Fourier Transform (FFT)

Discrete Fourier transform (DFT) $X[k]$ of an N-point signal $x[n]$ is given by

$$
\begin{cases}
X[k] = \displaystyle\sum_{n=0}^{N-1} x[n] W_N^{nk}, & k = 0, 1, \ldots, N-1 \\[2ex]
x[n] = \dfrac{1}{N} \displaystyle\sum_{n=0}^{N-1} X[k] W_N^{-nk}, & n = 0, 1, \ldots, N-1
\end{cases}
\tag{7.1}
$$

where $W_N = e^{-j2\pi/N}$. The above transform equations require N complex multiplications and $N-1$ complex additions for each term. For all N terms, N^2 complex multiplications and $N^2 - N$ complex additions are needed. As it is well known, the direct computation of (7.1) is not efficient.

To obtain a fast or real-time implementation of (7.1), one often uses a fast Fourier transform (FFT) algorithm, which makes use of the symmetry properties of DFT. There are many approaches to finding a fast implementation of DFT; that is, there are many variations of FFT algorithms. Here, we mention the approach presented in the *TI Application Report SPRA291* for computing a 2N-point FFT [2]. This approach involves forming two new N-point signals $x_1[n]$ and $x_2[n]$ from a 2N-point signal $g[n]$ by splitting it into an even and an odd part as follows:

$$
\begin{aligned}
x_1[n] &= g[2n] \qquad 0 \le n \le N-1 \\
x_2[n] &= g[2n+1]
\end{aligned}
\tag{7.2}
$$

From the two sequences $x_1[n]$ and $x_2[n]$, a new complex sequence $x[n]$ is defined to be

$$
x[n] = x_1[n] + jx_2[n] \qquad 0 \le n \le N-1
\tag{7.3}
$$

To get $G[k]$, the DFT of $g[n]$, the equation

$$
\begin{aligned}
G[k] &= X[k]A[k] + X^*[N-k]B[k] \\
k &= 0, 1, \ldots, N-1, \text{with } X[N] = X[0]
\end{aligned}
\tag{7.4}
$$

is used, where

$$
A[k] = \frac{1}{2}\left(1 - jW_{2N}^k\right)
\tag{7.5}
$$

and

$$
B[k] = \frac{1}{2}\left(1 + jW_{2N}^k\right)
\tag{7.6}
$$

Only N points of $G[k]$ are computed from (7.4). The remaining points are found by using the complex conjugate property of $G[k]$, that is, $G[2N-k]=G^*[k]$. As a result, a 2N-point transform is calculated based on an N-point transform, leading to a reduction in the number of operations.

7.2 Short-Time Fourier Transform (STFT)

Short-time Fourier transform (STFT) is a sequence of Fourier transforms of a windowed signal. STFT provides the time-localized frequency information for situations in which frequency components of a signal vary over time, whereas the standard Fourier transform provides the frequency information averaged over the entire signal time interval.

The STFT pair is given by

$$
\begin{cases}
\mathcal{X}_{STFT}[m, n] = \displaystyle\sum_{k=0}^{L-1} x[k]g[k-m]e^{-j2\pi nk/L} \\[4mm]
x[k] = \displaystyle\sum_{m}\sum_{n} \mathcal{X}_{STFT}[m, n]g[k-m]e^{j2\pi nk/L}
\end{cases} \tag{7.7}
$$

where $x[k]$ denotes a signal and $g[k]$ denotes an L-point window function. From (7.7), the STFT of $x[k]$ can be interpreted as the Fourier transform of the product $x[k]g[k-m]$. Figure 7-1 illustrates computing STFT by taking Fourier transforms of a windowed signal.

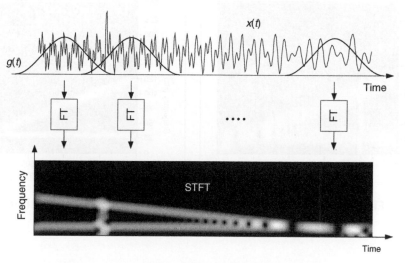

Figure 7-1: Short-time Fourier transform.

There exists a trade-off between time and frequency resolution in STFT. In other words, although a narrow-width window results in a better resolution in the time domain, it generates a poor resolution in the frequency domain, and vice versa. Visualization of STFT is often realized via its spectrogram, which is an intensity plot of STFT magnitude over time. Three spectrograms illustrating different time-frequency resolutions are shown in Figure 7-2. The implementation details of STFT are described in Lab 7.

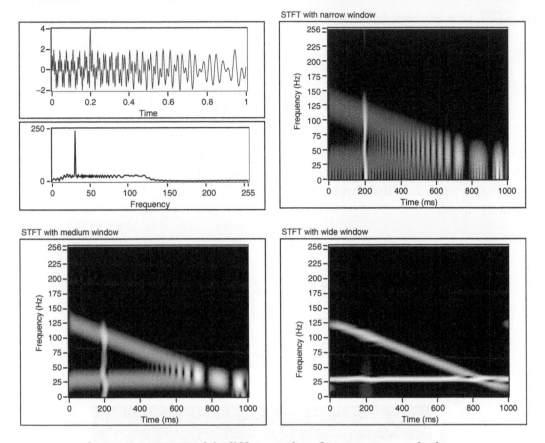

Figure 7-2: STFT with different time-frequency resolutions.

7.3 Discrete Wavelet Transform (DWT)

Wavelet transform offers a generalization of STFT. From a signal theory point of view, similar to DFT and STFT, wavelet transform can be viewed as the projection of a signal into a set of basis functions named wavelets. Such basis functions offer localization in the frequency domain. In contrast to STFT having equally spaced time-frequency localization, wavelet transform provides high frequency resolution at low frequencies and high time resolution at high frequencies. Figure 7-3 provides a tiling depiction of the time-frequency resolution of wavelet transform as compared to STFT and DFT.

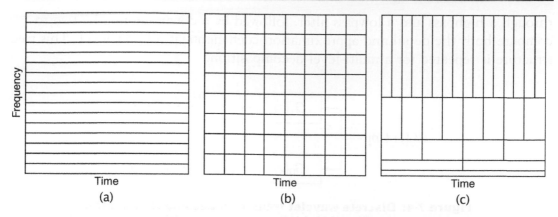

Figure 7-3: Time-frequency tiling for (a) DFT, (b) STFT, and (c) DWT.

The discrete wavelet transform (DWT) of a signal $x[n]$ is defined based on approximation coefficients, $W_\phi[j_0, k]$, and detail coefficients, $W_\psi[j, k]$, as follows:

$$W_\phi[j_0, k] = \frac{1}{\sqrt{M}} \sum_n x[n]\phi_{j_0,k}[n]$$

$$W_\psi[j, k] = \frac{1}{\sqrt{M}} \sum_n x[n]\psi_{j,k}[n] \qquad \text{for } j \geq j_0$$

(7.8)

and the inverse DWT is given by

$$x[n] = \frac{1}{\sqrt{M}} \sum_k W_\phi[j_0, k]\phi_{j_0,k}[n] + \frac{1}{\sqrt{M}} \sum_{j=j_0}^{J} \sum_k W_\psi[j, k]\psi_{j,k}[n]$$

(7.9)

where $n = 0, 1, 2, \ldots, M-1$, $j = 0, 1, 2, \ldots, J-1$, $k = 0, 1, 2, \ldots, 2^j-1$, and M denotes the number of samples to be transformed. This number is selected to be $M = 2^J$, where J indicates the number of transform levels. The basis functions $\{\phi_{j,k}[n]\}$ and $\{\psi_{j,k}[n]\}$ are defined as

$$\phi_{j,k}[n] = 2^{j/2}\phi[2^j n - k]$$

$$\psi_{j,k}[n] = 2^{j/2}\psi[2^j n - k]$$

(7.10)

where $\phi[n]$ is called the scaling function and $\psi[n]$ is called the wavelet function.

For an efficient implementation of DWT, the filter bank structure is often used. Figure 7-4 shows the decomposition or analysis filter bank for obtaining the forward DWT coefficients. The approximation coefficients at a higher level are passed through a highpass and a lowpass filter, followed by a downsampling by two to compute both the detail and approximation coefficients at a lower level. This tree structure is repeated for a multi-level decomposition.

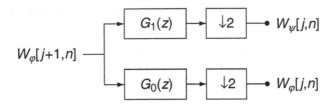

Figure 7-4: Discrete wavelet transform decomposition filter bank, G_0 lowpass and G_1 highpass decomposition filters.

Inverse DWT (IDWT) is obtained by using the reconstruction or synthesis filter bank shown in Figure 7-5. The coefficients at a lower level are upsampled by two and passed through a highpass and a lowpass filter. The results are added together to obtain the approximation coefficients at a higher level.

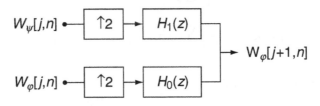

Figure 7-5: Discrete wavelet transform reconstruction filter bank, H_0 lowpass and H_1 highpass reconstruction filters.

7.4 Signal Processing Toolset

Signal Processing Toolset (SPT) is an add-on toolkit of LabVIEW that provides useful tools for performing time-frequency analysis [3]. SPT has three components: Joint Time-Frequency Analysis (JTFA), Super-Resolution Spectral Analysis (SRSA), and Wavelet Analysis.

The VIs associated with STFT are included as part of the JTFA component. The SRSA component is based on the model-based frequency analysis normally used for situations in which a limited number of samples is available. The VIs associated with the SRSA component include high-resolution spectral analysis and parameter estimation, such as amplitude, phase, damping factor, and damped sinusoidal estimation. The VIs associated with the Wavelet Analysis component include 1D and 2D wavelet transform as well as their filter bank implementations.

7.5 Bibliography

[1] C. Burrus, R. Gopinath, and H. Gao, *Wavelets and Wavelet Transforms A Primer*, Prentice-Hall, 1998.

[2] Texas Instruments, *TI Application Report SPRA291*.

[3] National Instruments, *Signal Processing Toolset User Manual*, Part Number 322142C-01, 2002.

Lab 7: FFT, STFT, and DWT

This lab shows how to use the LabVIEW tools to perform FFT, STFT, and DWT as part of a frequency domain transformation system.

L7.1 FFT Versus STFT

To illustrate the difference between FFT and STFT transformations, three signals are combined here to form a 512-point input signal: a 75 Hz sinusoidal signal sampled at 512 Hz, a chirp signal with linearly decreasing frequency from 200 to 120 Hz, and an impulse signal having an amplitude of 2 for 500 ms located at the 256th sample. This composite signal is shown in Figure L7-1. The FFT and STFT graphs are also shown in this figure. The FFT graph shows the time averaged spectrum reflecting the presence of a signal from 120 to 200 Hz, with one major peak at 75 Hz. As one can see from this graph, the impulse having the short time duration does not appear in the spectrum. The STFT graph shows the spectrogram for a time increment of 1 and a rectangular window of width 32 by which the presence of the impulse can be detected.

As far as the FP is concerned, two Menu Ring controls (**Controls » Modern » Ring & Enum » Menu Ring**) are used to input values via their labels. The labels and corresponding values of the ring controls can be modified by right-clicking and choosing **Edit Items**... from the shortcut menu. This brings up the dialog box shown in Figure L7-2.

An Enum (enumerate) control acts the same as a Menu Ring control, except that values of an Enum control cannot be modified and are assigned sequentially. A Menu Ring or Enum can be changed to a Ring Constant or Enum Constant when used on a BD.

Several spectrograms with different time window widths are shown in Figure L7-3. Figure L7-3(a) shows an impulse (vertical line) at time 500 ms because of the relatively time-localized characteristic of the window used. Even though a high resolution in the time domain is achieved with this window, the resolution in the frequency domain is so poor that the frequency contents of the sinusoidal and chirp signals cannot be easily distinguished. This is due to the Heisenberg's uncertainty principle [1], which states that if the time resolution is increased, the frequency resolution is decreased.

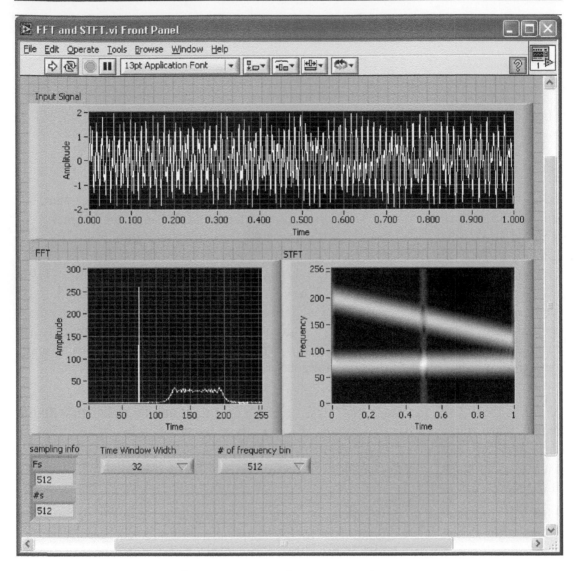

Figure L7-1: FP of FFT versus STFT.

Now, let us increase the width of the time-frequency window. This causes the frequency resolution to become better while the time resolution becomes poorer. As a result, as shown in Figure L7-3(d), the frequency contents of the sinusoidal and chirp signals become better distinguished. One can also see that as the time resolution becomes poorer, the moment of occurrence of the impulse becomes more difficult to identify.

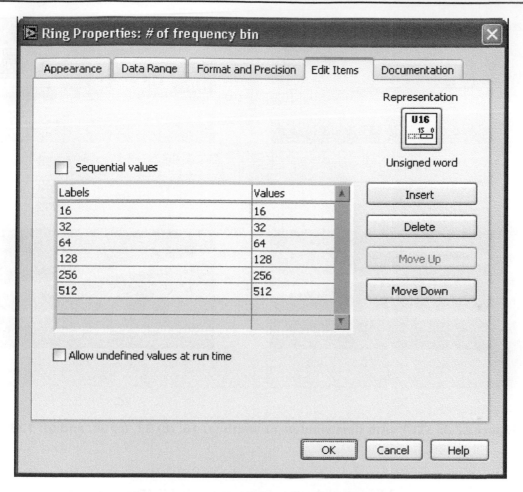

Figure L7-2: Properties of a ring control.

The BD of this example is illustrated in Figure L7-4. To build this VI, let us first generate the input signal with the specifications stated previously. Figure L7-5(a) shows the generation of the input signal (512 samples generated with the sampling frequency of 512 Hz) using a `MathScript Node`. In order to use this VI as the signal source of the system, an output terminal in the connector pane is wired to a waveform indicator. Then, the VI is saved as *Composite Signal.vi*.

Alternatively, the three signals can be generated using the built-in LabVIEW VIs and added together to form a composite signal; see Figure L7-5(b). The sinusoidal

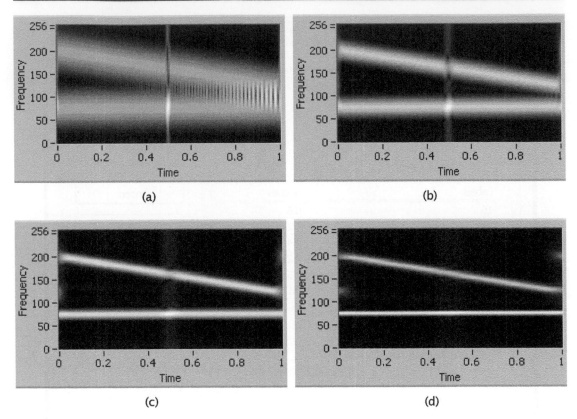

Figure L7-3: STFT with time window of width (a) 16, (b) 32, (c) 64, and (d) 128.

waveform is generated by using the `Sine Waveform` VI (**Functions » Signal Processing » Waveform Generation » Sine Waveform**), and the chirp signal is generated by using the `Chirp Pattern` VI (**Functions » Signal Processing » Signal Generation » Chirp Pattern**). Also, the impulse is generated by using the `Impulse Pattern` VI (**Functions » Signal Processing » Signal Generation » Impulse Pattern**).

Now, let us create the entire transformation system using the `Composite Signal` VI just made. Create a blank VI; then select **Functions » Select a VI....** This brings up a window for choosing and locating a VI. Click *Composite Signal.vi* to insert it into the BD. The composite signal output is connected to three blocks consisting of a waveform graph, an `FFT`, and an `STFT` VI. The waveform data (`Y` component) are connected to the input of the `FFT` VI (**Functions » Signal Processing » Transforms » FFT**). Only the first half of the output data from the `FFT` VI is taken, since the other half is a mirror image of the first half. This is done by placing an `Array Subset` function

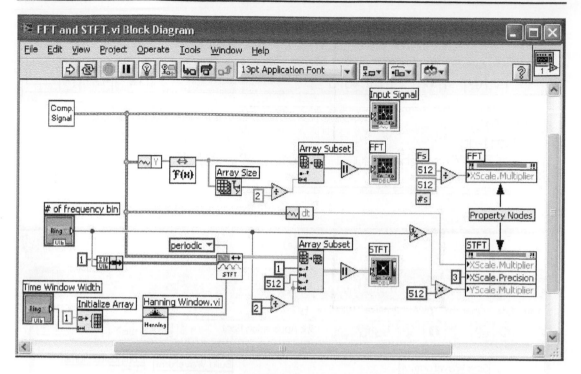

Figure L7-4: BD of FFT and STFT.

and wiring to it one half of the signal length. The magnitude of the FFT output is then displayed in the waveform graph. Properties of an FP object, such as scale multiplier of a graph, can be changed programmatically by using a property node. Property nodes are discussed in the next subsection.

Getting the STFT output is more involved than FFT. The STFT VI (**Functions » Addons » Time Frequency Analysis » Time Frequency Transform » STFT**), which is part of the Signal Processing Toolkit (SPT), is used here for this purpose. To utilize the STFT VI, one needs to connect several inputs as well as the input signal. These inputs are time-freq sampling info, extension, window info, and user-defined window. The time-freq sampling info is a cluster of time steps and frequency bins where time steps specify the sampling period along the time axis and the frequency bins indicate the FFT block size of the STFT. A constant of 1 is used for time steps in the example shown in Figure L7-4. The extension input specifies the method to pad data at both ends of a signal to avoid abrupt changes in the transformed outcome. There exist three different extension options: zero padding, symmetric, and periodic. The periodic mode is used in the example shown in Figure L7-4. The window info input specifies which commonly used sliding window to apply and

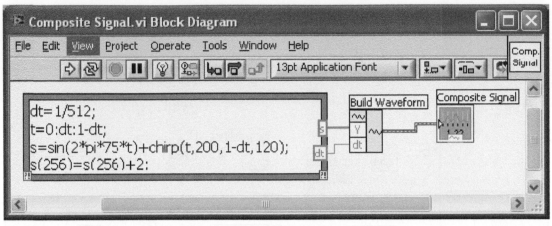

(a)

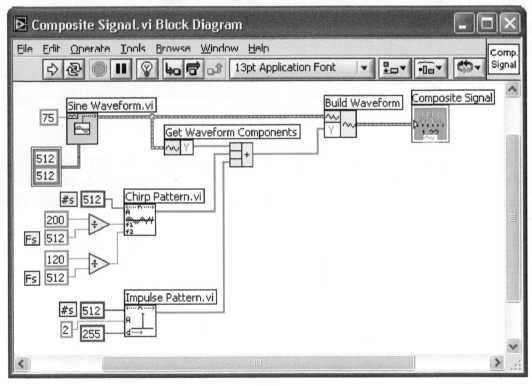

(b)

Figure L7-5: Composite signal (sine + chirp + impulse) generation using (a) MathScript Node and (b) graphical approach.

defines the resolution of the resulting time-frequency representation. On the other hand, the user-defined window input allows one to have a customized sliding window by specifying the coefficients. In our example, a Hanning window is considered by passing an array of all 1's whose width is adjustable by the user through the Hanning window VI (**Functions » Signal Processing » Window » Hanning Window**). Similar to FFT, only one-half of the frequency values are taken while the time values retain the original length. The start index of the array subset is set to one-half the number of frequency bins to access the positive frequency values, as shown in Figure L7-4. The reason is that the output of the STFT corresponding to the negative frequency values is followed by the output belonging to the positive frequency values. Additional details on using the STFT VI can be found in [2].

The output of the STFT is displayed in the Intensity Graph (**Controls » Modern » Graph » Intensity Graph**). Right-click on the Intensity Graph and then uncheck the **Loose Fit** option under both **X Scale** and **Y Scale** from the shortcut menu. When this is done, the STFT output graph gets fitted into the entire plotting area. Enable auto-scaling of intensity by right-clicking on the Intensity Graph and choosing **Z Scale » AutoScale Z**.

L7.1.1 Property Node

The number of FFT values varies based on the number of samples. Similarly, the number of frequency rows of STFT varies based on the number of frequency bins specified by the user. However, the scale of the frequency axis in FFT or STFT graphs should always remain between 0 and $f_s/2$, which is 256 Hz in the example, regardless of the number of frequency bins, as illustrated in Figure L7-1 and Figure L7-3. For this reason, the multiplier for the spectrogram scale needs to be changed depending on the width of the time window during run time.

A property node can be used to modify the appearance of an FP object. A property node can be created by right-clicking either on a terminal icon in a BD or an object in an FP, and then by choosing the **visible** property element through **Create » Property Node**. This way, the default element of the chosen property gets created in a BD, which is linked to a corresponding FP object. Various property elements of the property node can be modified to reflect the read or the write mode. Note that, by default, a property node is set to read. To change to the write mode, right-click on a property element and choose **Change to Write**. The read/write mode of all elements can be changed together by choosing **Change all to Read/Write**.

To change the scale of the spectrogram graph, one needs to modify the value of the element **YScale.Multiplier**. Replace the element **visible** with **YScale.Multiplier** by clicking it and choosing **Y Scale » Offset and Multiplier » Multiplier**. The sampling frequency of the signal divided by the number of frequency bins, which defines the scale multiplier, is wired to the element **YScale.Multiplier** of the property node. Two more elements, **XScale.Multiplier** and **XScale.Precision**, are added to the property node for modifying the time axis multiplier and precision, respectively.

A property node of the FFT graph is also created and modified in a similar way considering that the resolution of FFT is altered depending on the sampling frequency and number of input signal samples. The property nodes of the STFT and FFT graphs are shown in Figure L7-4. More details on using property nodes can be found in [3].

L7.2 DWT

In this transformation, the time-frequency window has high frequency resolution for higher frequencies and high time resolution for lower frequencies. This is a great advantage over STFT where the window size is fixed for all frequencies.

The BD of a 1D decomposition and reconstruction wavelet transform is shown in Figure L7-6. Three VIs including WA Wavelet Filter VI (**Functions » Addons » Wavelet Analysis » Discrete Wavelet » Filter Banks**), WA Discrete Wavelet

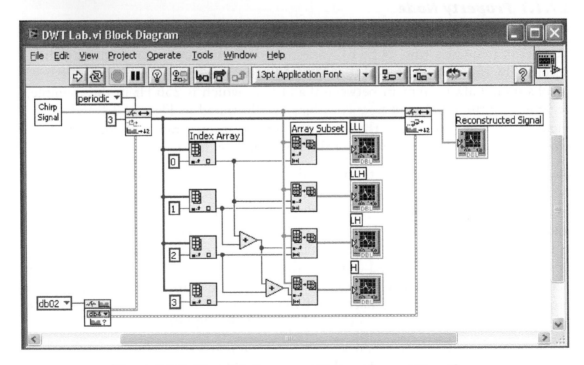

Figure L7-6: Wavelet decomposition and reconstruction.

Transform VI, and WA Inverse Discrete Wavelet Transform VI
(**Functions » Addons » Wavelet Analysis » Discrete Wavelet**) are used here from the
wavelet analysis palette.

A chirp type signal, shown in Figure L7-7, is considered to be the input signal source.
This signal is designed to consist of four sinusoidal signals, each consisting of 128
samples with increasing frequencies in this order: 250, 500, 1000, 2000 Hz. This
makes the entire chirp signal 512 samples. The Fourier transform of this signal is also
shown in Figure L7-7.

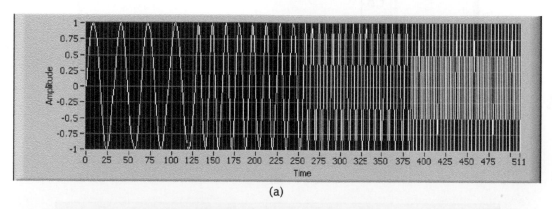

(a)

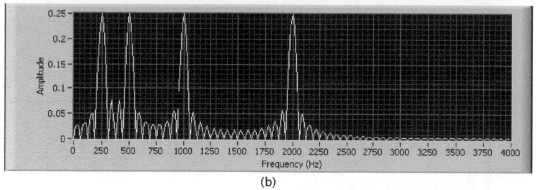

(b)

Figure L7-7: Waveforms of input signal: (a) time domain and (b) frequency domain.

Figure L7-8(a) illustrates the BD of this signal generation process. Save this VI as *Chirp
Signal.vi* to be used as a signal source subVI within the DWT VI. Note that the **Concatenate
Inputs** option of the Build Array function should be chosen to build the 1D chirp
signal. This VI has only one output terminal. As an alternative to the graphical approach,
a MATLAB Script Node can be used to generate the chirp signal. This way, the four
signals need to be concatenated using the operator [ ], as shown in Figure L7-8(b).

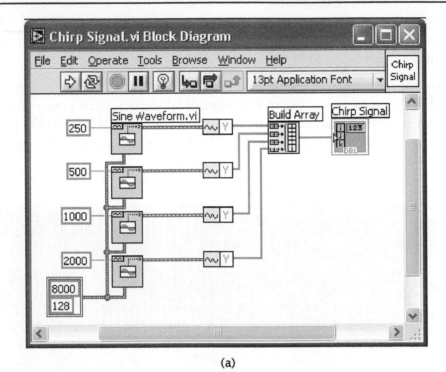

(a)

(b)

Figure L7-8: Generating input signal using (a) graphical approach and (b) textual approach.

The WA Discrete Wavelet Transform VI requires four inputs, including input signal, extension, levels, and analysis filter. The input signal is provided by the Chirp Signal VI. For the extension input, the same options are available as mentioned earlier for STFT. The input levels specify the number of levels of decomposition. In the BD shown in Figure L7-6, a three-level decomposition is used via specifying a constant 3. The filter bank implementation for a three-level wavelet decomposition is illustrated in Figure L7-9. In this example, the Daubechies-2 wavelet is used. The coefficients of the filters are generated by the Wavelet Filter VI. This VI provides the coefficient sets for both the decomposition and reconstruction parts.

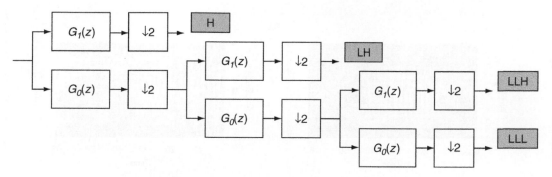

Figure L7-9: Waveform decomposition tree.

The result of the WA Discrete Wavelet Transform VI is structured into a 1D array corresponding to the components of the transformed signal in the order LLL, LLH, LH, H, where L stands for low and H for high. The length of each component is also available from this VI. The wavelet decomposed outcome for each stage of the filter bank is shown in Figure L7-10. From the outcome, it can be observed that lower frequencies occur earlier and higher frequencies occur later in time. This demonstrates the fact that wavelet transform provides both frequency and time resolution, a clear advantage over Fourier transform.

The decomposed signal can be reconstructed by the WA Inverse Discrete Wavelet Transform VI. From the reconstructed signal, shown in Figure L7-10, one can see that the wavelet decomposed signal is reconstructed perfectly by using the synthesis or reconstruction filter bank.

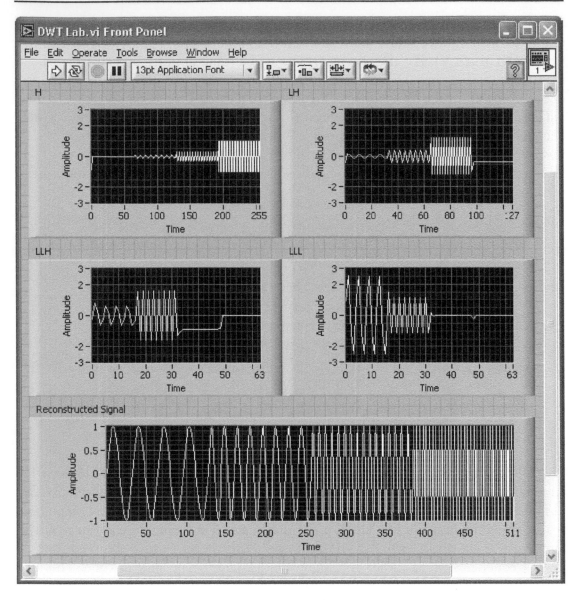

Figure L7-10: FP of DWT VI.

L7.3 Bibliography

[1] C. Burrus, R. Gopinath, and H. Gao, *Wavelets and Wavelet Transforms—A Primer*, Prentice-Hall, 1998.

[2] National Instruments, *Signal Processing Toolset User Manual*, Part Number 322142C-01, 2002.

[3] National Instruments, *LabVIEW User Manual*, Part Number 320999E-01, 2003.

L7.4 Lab Experiments

1. Build a VI to show the difference between the FFT and STFT transformations on a 512-point composite signal consisting of the following four components added together: (i) a 215 Hz sinusoidal signal sampled at 1200 Hz, (ii) a 125 Hz sinusoidal signal sampled at 1200 Hz with a phase shift of 90 degrees, (iii) a chirp signal sampled at 1200 Hz with linearly increasing frequency from 150 Hz to 200 Hz, and (iv) an impulse signal having an amplitude of 5 at the 75th and 240th samples. Generate the composite signal with and without using the MathScript feature. For each case, compute and display both FFT and STFT. Observe the STFT output for varying levels of frequency bins and time window widths. Also, compare the STFT output for different user-defined windows such as Hanning and Hamming windows.

2. Build a VI to show the difference between the FFT and STFT transformations on a 512-point composite signal consisting of the following four components added together: (i) a 175 Hz sinusoidal signal sampled at 1500 Hz, (ii) a 225 Hz sinusoidal signal sampled at 1500 Hz with a phase shift of 90 degrees, (iii) a chirp signal sampled at 1500 Hz with linearly increasing frequency from 350–425 Hz, and (iv) an impulse signal having an amplitude of 10 at the 165th and 235th samples. Generate the composite signal with and without using the MathScript feature. For each case, compute and display both FFT and STFT. Observe the STFT output for varying levels of frequency bins and time window widths. Also, compare the STFT output for different user-defined windows such as Chebyshev and Kaiser-Bessel windows.

3. Build a VI to show the four-level DWT transformation on a 512-point composite signal consisting of four components each having 128 samples, namely three sinusoidal signals with increasing frequencies 125 Hz, 235Hz, 455Hz, and a chirp signal with linearly increasing frequency from 500 Hz to 700 Hz. Use a sampling

frequency of 8000 Hz. Generate the composite signal with and without using the MathScript feature. For each case, compute and display the four-level wavelet decomposition and reconstruction of the composite signal.

4. Build a VI to show the five-level DWT transformation on a 512-point composite signal consisting of four components each having 128 samples, namely three sinusoidal signals with increasing frequencies 165 Hz, 295Hz, 575Hz, and a chirp signal with linearly increasing frequency from 700 Hz to 900 Hz. Use a sampling frequency of 8000 Hz. Generate the composite signal with and without using the MathScript feature. Also, extract the wavelet decomposed outcomes for each stage of the filter bank with and without using the MathScript feature. For each case, compute and display the five-level wavelet decomposition and reconstruction of the composite signal.

CHAPTER 8

DSP Implementation Platform: TMS320C6x Architecture and Software Tools

Implementing some or most components of a signal processing system on a DSP processor is often computationally more efficient. The choice of a DSP processor to use in a signal processing system is application dependent. Many factors influence this choice, including cost, performance, power consumption, ease-of-use, time-to-market, and integration/interfacing capabilities.

8.1 TMS320C6X DSP

The family of TMS320C6x processors, manufactured by Texas Instruments, is built to deliver speed. They are designed for a million instructions per second (MIPS) intensive applications such as digital video. There are many processor versions belonging to this family differing in instruction cycle time, speed, power consumption, memory, peripherals, packaging, and cost. For example, the fixed-point C6416-600 version operates at 600 MHz (1.67 ns cycle time), delivering a peak performance of 4800 MIPS. The floating-point C6713-225 version operates at 225 MHz (4.4 ns cycle time), delivering a peak performance of 1350 MIPS.

Figure 8-1 shows a block diagram of the generic C6x architecture. The C6x central processing unit (CPU) consists of eight functional units divided into two sides: A and B. Each side has an .M unit (used for multiplication operation), an .L unit (used for logical and arithmetic operations), an .S unit (used for branch, bit manipulation, and arithmetic operations), and a .D unit (used for loading, storing, and arithmetic operations). Some instructions such as ADD can be done by more than one unit. Sixteen 32-bit registers are associated with each side. Interaction with the CPU must be done through these registers.

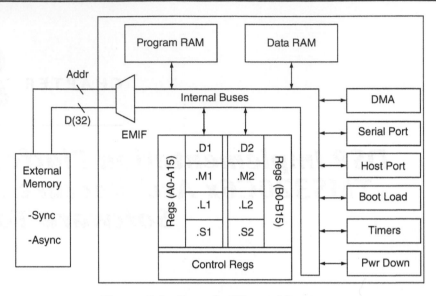

Figure 8-1: Generic C6x architecture.

As shown in Figure 8-2 the internal buses consist of a 32-bit program address bus, a 256-bit program data bus accommodating eight 32-bit instructions, two 32-bit data address buses (DA1 and DA2), two 32-bit (64-bit for C64 version) load data buses (LD1 and LD2), and two 32-bit (64-bit for the floating-point version) store data buses (ST1 and ST2). In addition, there are a 32-bit direct memory access (DMA) data and a 32-bit DMA address bus. The off-chip, or external, memory is accessed through a 20-bit address bus and a 32-bit data bus.

The peripherals on a typical C6x processor include External Memory Interface (EMIF), DMA, Boot Loader, Multi-channel Buffered Serial Port (McBSP), Host Port Interface (HPI), Timer, and Power Down unit. EMIF provides the necessary timing for accessing external memory. DMA allows the movement of data from one place in memory to another place without interfering with the CPU operation. Boot Loader boots the code from off-chip memory or HPI to the internal memory. McBSP provides a high-speed multi-channel serial communication link. HPI allows a host to access the internal memory. Timer provides two 32-bit counters. The Power Down unit is used to save power for durations when the CPU is inactive.

8.1.1 Pipelined CPU

In general, there are three basic steps to perform an instruction. They include fetching, decoding, and execution. If these steps are done serially, not all of the resources on the processor, such as multiple buses or functional units, are fully utilized. In order to increase throughput, DSP CPUs are designed to be pipelined.

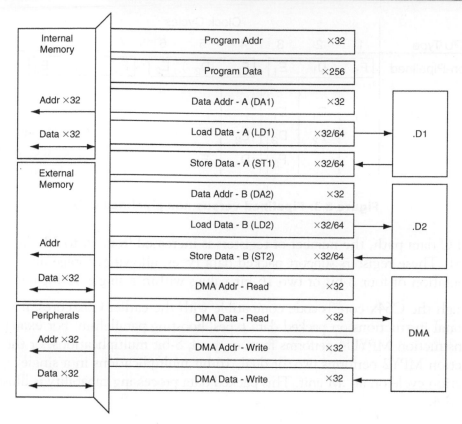

Figure 8-2: C6x internal buses.

This means that the foregoing steps are carried out simultaneously. Figure 8-3 illustrates the difference in the processing time for three instructions executed on a serial or non-pipelined and a pipelined CPU. As one can see from this figure, a pipelined CPU requires fewer clock cycles to complete the same number of instructions.

The C6x architecture is based on the Very Long Instruction Word (VLIW) architecture. In such architectures, several instructions are captured and processed simultaneously. For more details on the TMS320C6000 architecture, the interested reader is referred to [1].

8.1.2 C64x DSP

The C64x is a more recently released DSP core, as part of the C6x family, with higher MIPS power operating at higher clock rates. This core can operate in the range of 300–1000 MHz clock rates, giving a processing power of 2400–8000 MIPS. The C64x speedups are achieved due to many enhancements, some of which are mentioned here.

	Clock Cycles								
CPU Type	1	2	3	4	5	6	7	8	9
Non-Pipelined	F_1	D_1	E_1	F_2	D_2	E_2	F_3	D_3	E_3

Pipelined

F_1 D_1 E_1
F_2 D_2 E_2
F_3 D_3 E_3

F_x = fetching of instruction x
D_x = decoding of instruction x
E_x = execution of instruction x

Figure 8-3: Pipelined versus non-pipelined CPU.

Per CPU data path, the number of registers is increased from 16 to 32, A0–A31 and B0–B31. These registers support packed data types, allowing storage and manipulation of four 8-bit or two 16-bit values within a single 32-bit register.

Although the C64x core is code compatible with the earlier C6x cores, it can run additional instructions on packed data types, boosting parallelism. For example, the new instruction MPYU4 performs four, or quad, 8-bit multiplications, or the instruction MPY2 performs two, or dual, 16-bit multiplications in a single instruction cycle on an .M unit. This packed data processing capability is illustrated in Figure 8-4.

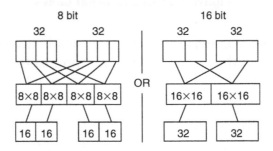

Figure 8-4: C64x packed data processing capability.

Additional instructions have been added to each functional unit on the C64x for performing special-purpose functions encountered in wireless and digital imaging applications. In addition, the functionality of each functional unit on the C64x has been improved, leading to a greater orthogonality, or generality, of operations. For example, the .D unit can perform 32-bit logical operation just as the .S and .L units, or the .M unit can perform shift and rotate operations just as the .S unit.

8.2 C6x DSK Target Boards

Upon the availability of a DSP Starter Kit (DSK) board, appropriate components of a DSP system can be run on an actual C6x processor.

8.2.1 Board Configuration and Peripherals

As shown in Figure 8-5, the C6713 DSK board is a DSP platform which includes a C6713 floating-point DSP chip operating at 225 MHz with 8 Mbytes of on-board synchronous dynamic RAM (SDRAM), 512 Kbytes of flash memory, and a 16-bit stereo codec AIC23. The codec is used to convert an analog input signal to a digital signal for the DSP manipulation. The sampling frequency of the codec can be changed from 8 kHz to 96 kHz. The C6416 DSK board includes a C6416 fixed-point DSP chip operating at 600 MHz with 16 Mbytes of on-board SDRAM, 512 Kbytes of flash memory, and an AIC23 codec.

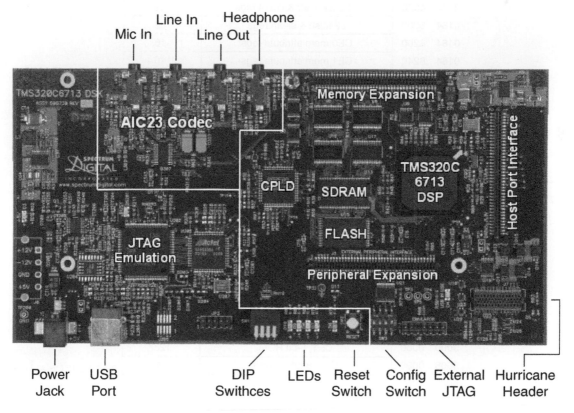

Figure 8-5: C6713 DSK board [2].

8.2.2 *Memory Organization*

The external memory used by a DSP processor can be either static or dynamic. Static memory (SRAM) is faster than dynamic memory (DRAM), but it is more expensive because it takes more space on silicon. SDRAM (synchronous DRAM) provides a compromise between cost and performance.

Given that the address bus is 32 bits wide, the total memory space of C6x consists of $2^{32} = 4$ Gbytes. For the lab exercises in this book, the DSK board is configured based on the memory map 1 shown in Figure 8-6.

Address	Memory Map 1	Block Size (Bytes)
0000 0000	Internal RAM (L2)	64K
0001 0000	Reserved	24M
0180 0000	EMIF control regs	32
0184 0000	Cache Configuration regs	4
0184 4000	L2 base addr & count regs	32
0184 4020	L1 base addr & count regs	32
0184 5000	L2 flush & clean regs	32
0184 8200	CE0 mem attribute regs	16
0184 8240	CE1 mem attribute regs	16
0184 8280	CE2 mem attribute regs	16
0184 82C0	CE3 mem attribute regs	16
0188 0000	HPI control regs	4
018C 0000	McBSP0 regs	40
0190 0000	McBSP1 regs	40
0194 0000	Timer0 regs	12
0198 0000	Timer1 regs	12
019C 0000	Interrupt selector regs	12
01A0 0000	EDMA parameter RAM	2M
01A0 FFE0	EDMA control regs	32
0200 0000	QDMA regs	20
0200 0020	QDMA pseudo-regs	20
3000 0000	McBSP0 data	64M
3400 0000	McBSP1 data	64M
8000 0000	CE0, SDRAM	16M
9000 0000	CE1, 8-bit ROM	128K
9008 0000	CE1, 8-bit I/O port	4
A000 0000	CE2-Daughtercard	256M
B000 0000	CE3-Daughtercard	256M
10000 0000		

Figure 8-6: C6x DSK memory map [3].

If a program fits into the on-chip or internal memory, it should be run from there to avoid delays associated with accessing off-chip or external memory. If a program is too big to be fitted into the internal memory, most of its time-consuming portions should be placed into the internal memory for efficient execution. For repetitive codes, it is recommended that the internal memory is configured as cache memory. This allows accessing external memory as seldom as possible and hence avoiding delays associated with such accesses.

8.3 DSP Programming

Most DSP processors can be programmed either in C or assembly. Although writing programs in C would require less effort, the efficiency achieved is normally less than that of programs written in assembly. Efficiency means having as few instructions or as few instruction cycles as possible by making maximum use of the resources on the chip.

In practice, one starts with C coding to analyze the behavior and functionality of an algorithm. Then, if the required processing time is not met by using the C compiler optimizer, the time-consuming portions of the C code are identified and converted into assembly, or the entire code is rewritten in assembly. In addition to C and assembly, the C6x allows writing code in linear assembly. Figure 8-7 illustrates code efficiency versus coding effort for three types of source files on the C6x: C, linear assembly, and hand-optimized assembly. As one can see, linear assembly provides a good compromise between code efficiency and coding effort.

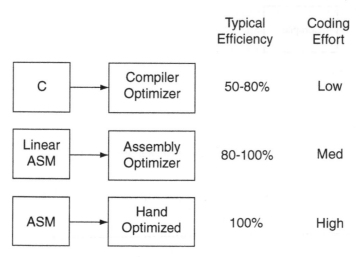

Figure 8-7: Code efficiency versus coding effort [1].

More efficient codes are obtained by performing assembly programming fully utilizing the pipelined feature of the CPU. Details regarding programming in assembly/linear assembly and code optimization are discussed in [1].

8.3.1 Software Tools: Code Composer Studio

The assembler is used to convert an assembly file into an object file (.obj extension). The assembly optimizer and the compiler are used to convert, respectively, a linear assembly file and a C file into an object file. The linker is used to combine object files, as instructed by the linker command file (.cmd extension), into an executable file. All the assembling, linking, compiling, and debugging steps have been incorporated into an integrated development environment (IDE) called Code Composer Studio (CCS or CCStudio). CCS provides an easy-to-use graphical user environment for building and debugging C and assembly codes on various target DSPs. Figure 8-8 shows the steps involved in going from a source file (.c extension for C, .asm for assembly, and .sa for linear assembly) to an executable file (.out extension).

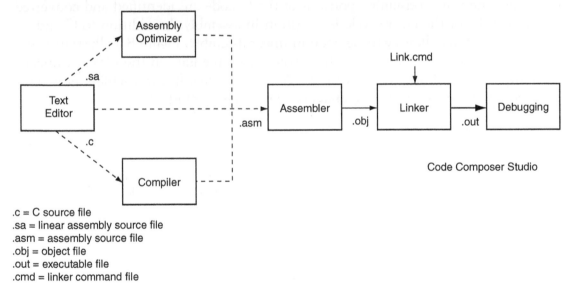

```
.c = C source file
.sa = linear assembly source file
.asm = assembly source file
.obj = object file
.out = executable file
.cmd = linker command file
```

Figure 8-8: C6x software tools [1].

During its setup, CCS can be configured for different target DSP boards (e.g., C6713 DSK, C6416 DSK, C6xxx Simulator). The version used in the book is CCS 3.0. CCS provides a file management environment for building application programs. It includes an integrated editor for editing C and assembly files. For debugging purposes, it provides breakpoints, data monitoring and graphing capabilities, profiler for benchmarking, and probe points to stream data to and from a target DSP.

8.3.2 Linking

Linking places code, constant, and variable sections into appropriate locations in memory as specified in a linker command file. Also, it combines several object files into a final executable file. A typical command file corresponding to the DSK memory map 1 is shown in Figure 8-9.

```
MEMORY
{
        VECS:   o=00000000h   l=00000200h    /* interrupt vectors */
        PMEM:   o=00000200h   l=0000FE00h    /* Internal RAM (L2) mem */
        BMEM:   o=80000000h   l=01000000h    /* External Memory CE0, SDRAM, 16
Mbytes */
}

SECTIONS
{
    .intvecs    >  0h
    .text       >  PMEM
    .far        >  PMEM
    .stack      >  PMEM
    .bss        >  PMEM
    .cinit      >  PMEM
    .pinit      >  PMEM
    .cio        >  PMEM
    .const      >  PMEM
    .data       >  PMEM
    .switch     >  PMEM
    .sysmem     >  PMEM
}
```

Figure 8-9: A typical command file.

The first part, MEMORY, provides a description of the type of physical memory, its origin and its length. The second part, SECTIONS, specifies the assignment of various code sections to the available physical memory. These sections are defined by directives such as .text, .data, etc.

8.3.3 Compiling

The build feature of CCS can be used to perform the entire process of compiling, assembling, and linking in one step. The compiler allows four levels of optimizations.

Debugging and full-scale optimization cannot be done together, since they are in conflict; that is, in debugging, information is added to enhance the debugging process, whereas in optimizing, information is minimized or removed to enhance code efficiency. In essence, the optimizer changes the flow of C code, making program debugging very difficult.

It is thus a good programming practice to first verify the proper functionality of codes by using the compiler with no optimization. Then, use full optimization to make them efficient. It is recommended that an intermediary step be taken in which some optimization is done without interfering with source-level debugging. This intermediary step can re-verify code functionality before performing full optimization.

8.4 Bibliography

[1] N. Kehtarnavaz, *Real-Time Digital Signal Processing Based on the TMS320C6000*, Elsevier, 2005.

[2] Texas Instruments "C6713 DSK Tutorial," *Code Composer Studio 2.2 C6713 DSK Help*, 2003.

[3] Texas Instruments, *TMS320C6711, TMS320C6711B, TMS320C6711C, TMS320C6711D Floating-Point Digital Signal Processors*, SPRS088L, 2004.

Lab 8: Getting Familiar with Code Composer Studio

L8.1 Code Composer Studio

This tutorial lab introduces the basic features of CCS one needs to know in order to build and debug a C program running on a DSP processor. To become familiar with all of its features, refer to the *TI CCS Tutorial* [1] and *TI CCS User's Guide* manuals [2].

This lab demonstrates how a simple DSP program can be compiled and linked by using CCS. The algorithm consists of a sinewave generator via using a difference equation. As part of this example, debugging and benchmarking issues are also covered. Knowledge of C programming is required for this and the next lab. These labs may be skipped if the reader is interested only in the LabVIEW implementation.

Note that the accompanying CD provides the lab programs separately for the DSP target boards C6416 and C6713 DSK, as well as the simulator.

L8.2 Creating Projects

Let us consider all the files required to create an executable file; that is, `.c` (c), `.asm` (assembly), and `.sa` (linear assembly) source files; a `.cmd` linker command file; an `.h` header file; and appropriate `.lib` library files. The CCS code development process begins with the creation of a *project* to easily integrate and manage all these required files for generating and running an executable file. After opening CCS by double-clicking the CCS icon on the desktop, one can see the **Project View** panel on the left side of the CCS window. This panel provides an easy mechanism for building a project. In this panel, a project file (`.pjt` extension) can be created or opened to contain not only the source and library files, but also the compiler, assembler, and linker options for generating an executable file.

To create a project, choose the menu item **Project » New** from the CCS menu bar. This brings up the dialog box **Project Creation**, as shown in Figure L8-1. In the dialog box, navigate to the working folder, here considered to be C:\CCStudio\ *myprojects*, and type a project name in the field **Project Name**. Then, click the button **Finish** for CCS to create a project file named *Lab08.pjt*. All the files necessary to run a program should be added to the project.

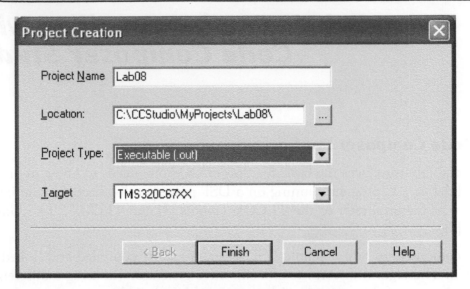

Figure L8-1: Creating a new project.

CCS provides an integrated editor which allows the creation of source files. Some of the features of the editor are color syntax highlighting, code block marking in parentheses and braces, parenthesis/brace matching, control indentions, and find/replace/search capabilities. It is also possible to add files to a project from Windows Explorer using the drag-and-drop approach. An editor window is brought up by choosing the menu item **File » New » Source File**. For this lab, let us type the following C code into the editor window:

```c
#include <stdio.h>
#include <math.h>

#define  PI           3.141592

void main()
{
    short  i, gain;
    float fs, f;
    float y[3], a[2], b1, x;

    short *output;
    output = (short *) 0x0000FF00;

    // Coefficient Initialization

    fs = 8000;                          // Sampling frequency
    f = 500;                            // Signal frequency
    gain = 100;                         // Amplitude gain
```

```
        a[0] = -1;
        a[1] = 2 * cos(2*PI*f/fs);
        b1 = gain;

        // Initial Conditions

        y[1] = y[2] = 0;
        x = 1;

        printf("BEGIN\n");

        for (i=0; i<128; i++)
        {
                y[0] = b1*x + a[1]*y[1] + a[0]*y[2];
                y[2] = y[1];
                y[1] = y[0];
                x = 0;

                output[i] = (short) y[0];
        }
        printf("END\n");
}
```

This code generates a sinusoidal waveform $y[n]$ based on the following difference equation:

$$y[n] = B_1 x[n-1] + A_1 y[n-1] + A_0 y[n-2] \qquad (8.1)$$

where $B_1 = 1$, $A_0 = -1$, $A_1 = 2\cos(\theta)$, and $x[n]$ is a delta function. The frequency of the waveform is given by [1]

$$f = \frac{f_s}{2\pi} \arccos(A_1/2) \qquad (8.2)$$

By changing the coefficient A_1, one can alter the frequency. By changing the coefficient B_1, one can alter the gain.

Save the created source file by choosing the menu item **File » Save**. This brings up the dialog box **Save As**, as shown in Figure L8-2. In the dialog box, go to the field **Save as type** and select **C Source Files (*.c)** from the pull-down list. Then, go to the field **File name** and type **sinewave.c**. Finally, click **Save** to save the code into a C source file named *sinewave.c*.

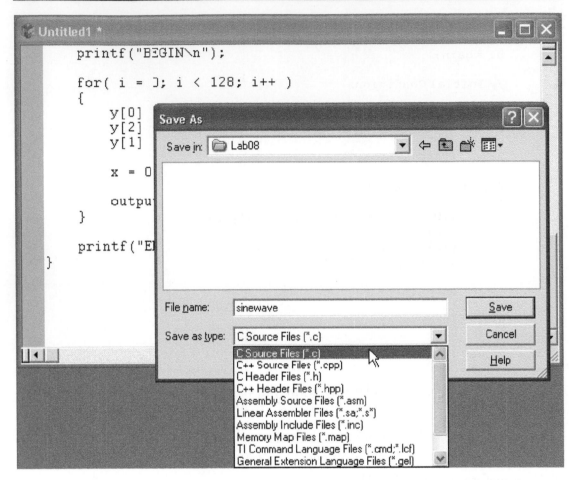

Figure L8-2: Creating a source file.

In addition to source files, a linker command file must be specified to create an executable file and to conform to the memory map of the target DSP. A linker command file can be created by choosing **File » New » Source File**. For this lab, let us type the command file shown in Figure L8-3. This file can also be downloaded from the accompanying CD, which is configured based on the DSK memory map. Save the editor window into a linker command file by choosing **File » Save** or by pressing <Ctrl-S>. This brings up the dialog box **Save As**. Go to the field **Save as type** and select **TI Command Language Files (*.cmd)** from the pull-down list. Then, type Lab08.cmd in the field **File name** and click **Save**.

```
MEMORY
{
        PMEM:   o=00000000h    1=0000FF00h    /* Internal RAM (L2) mem    */
        PMEM2:  o=0000FF00h    1=00000100h    /* Defined for output value*/
        BMEM:   o=80000000h    1=01000000h    /* CE0, SDRAM, 16 MBytes    */
}

SECTIONS
{
        .text       >       PMEM
        .far        >       PMEM
        .stack      >       PMEM
        .bss        >       PMEM
        .cinit      >       PMEM
        .pinit      >       PMEM
        .cio        >       PMEM
        .const      >       PMEM
        .data       >       PMEM
        .switch     >       PMEM
        .sysmem     >       PMEM
}}
```

Figure L8-3: Linker command file for Lab 8.

Now that the source file *sinewave.c* and the linker command file *Lab08.cmd* are created, they should be added to the project for compiling and linking. To do this, choose the menu item **Project » Add Files to Project**. This brings up the dialog box **Add Files to Project**. In the dialog box, select **sinewave.c** and click the button **Open**. This adds *sinewave.c* to the project. In order to add *Lab08.cmd*, choose **Project » Add Files to Project**. Then, in the dialog box **Add Files to Project**, set **Files of type** to **Linker Command File (*.cmd)** so that *Lab08.cmd* appears in the dialog box. Next, select **Lab08.cmd** and click the button **Open**. In addition to *sinewave.c* and *Lab08.cmd* files, the run-time support library should be added to the project. To do so, choose **Project » Add Files to Project**, go to the compiler library folder (here assumed to be the default option *C:\CCStudio\c6000\cgtools\lib*), select **Object and Library Files (*.o*,*.l*)** in the box **Files of type**, then select *rts6700.lib*, and click **Open**. If running on the TMS320C6416, select *rts6400.lib* instead.

After one adds all the source files, the command file, and the library file to the project, it is time to either build the project or to create an executable file for the target DSP. This is achieved by choosing the **Project » Build** menu item. CCS compiles, assembles, and links all of the files in the project. Messages about this process are shown in a panel at the bottom of the CCS window. When the building process is completed without any errors, the executable file *Lab08.out* is generated. It is also possible to do incremental builds—that is, rebuilding only those files changed since last build—by choosing the menu item **Project » Rebuild**. The CCS window provides shortcut buttons for frequently used menu options, such as **Incremental Build** and **Rebuild All**.

Although CCS provides default build options, all the compiler, assembler, and linker options can be changed via the menu item **Project » Build Options**. Among many compiler options shown in Figure L8-4, particular attention should be paid to the optimization level options. There are four levels of optimization (0, 1, 2, 3), which control the type and degree of optimization. Note that in some cases, debugging is not possible due to optimization. Thus, it is recommended to first debug a program to make sure that it is logically correct before performing any optimization.

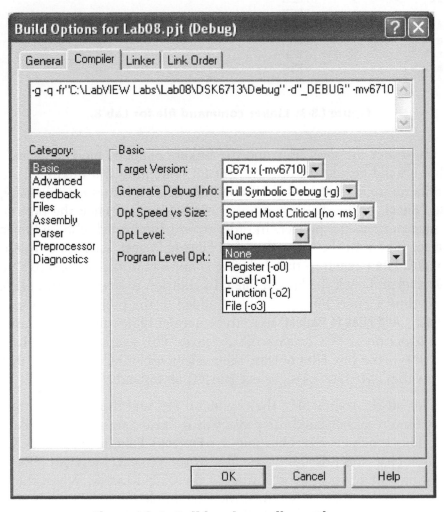

Figure L8-4: Build and compiler options.

Another important compiler option is the **Target Version** option. When implementing on the floating-point target DSP (TMS320C6713 DSK), go to the **Target Version** field and select **C671x (-mv6710)** from the pull-down list. For the fixed-point target DSP (TMS320C6416 DSK), select **C64xx (-mv6400)**.

When a message stating a compilation error appears, click **Stop Build** and scroll up in the build area to see the syntax error message. Double-click on the red text that describes the location of the syntax error. Notice that the *sinewave.c* file opens, and the cursor appears on the line that has caused the error, as shown in Figure L8-5. After one corrects the syntax error, the file should be saved and the project rebuilt.

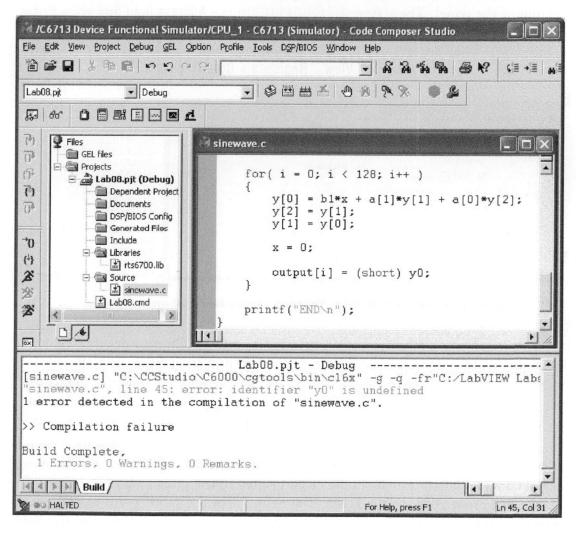

Figure L8-5: Build error.

L8.3 Debugging Tools

Once the build process is completed without any errors, the program can be loaded and executed on the target DSP. To load the program, choose **File » Load Program**, select the program *Lab08.out* just rebuilt, and click **Open**. To run the program, choose the menu item **Debug » Run**. One should be able to see BEGIN and END appearing in the **Stdout** window.

Now, let us view the content of the memory at a specific location. To do so, select **View » Memory** from the menu. The **Memory Window** panel should appear on the right side of the CCS window. Select **16-bit Signed Int** from the pull-down list at the bottom of the panel. Then, type 0x0000FF00 in the **Address** field and press Enter. A memory window grid should appear in the middle of the panel, as shown in Figure L8-6. The contents of the CPU, peripheral, DMA, and serial port registers can also be viewed by selecting **View » Registers » Core Registers**.

Figure L8-6: Memory Window Options dialog box and memory window.

Data stored in the DSP memory can be saved to a file. CCS provides a probe point capability so that a stream of data can be moved from the DSP to a PC host file or vice versa. In order to use this capability, one should set a probe point within the program by placing a mouse cursor at the line where a stream of data needs to be transferred and by clicking the button **Probe Point** 🔍 (see Figure L8-7). Choose **File » File I/O** to invoke the dialog box **File I/O**, as shown in Figure L8-8. Select the tab **File Output** for saving the data file; then click the button **Add File** and type the name of the data file. Next, the file should be connected to the probe point by clicking the button **Add Probe Point**. In the **Probe Point** field, select the probe point to make it active; then connect the probe point to the PC file through **File Out:...** in the field **Connect To**. Click the button **Replace** and then the button **OK**.

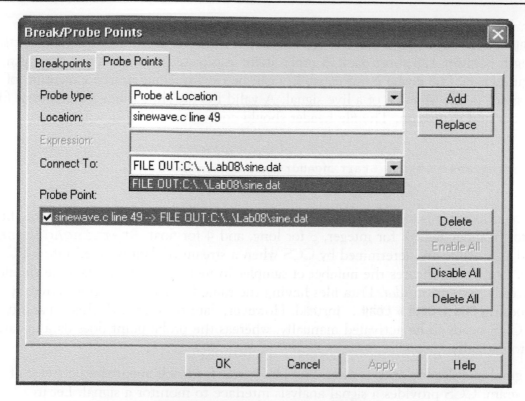

Figure L8-7: Probe Points window.

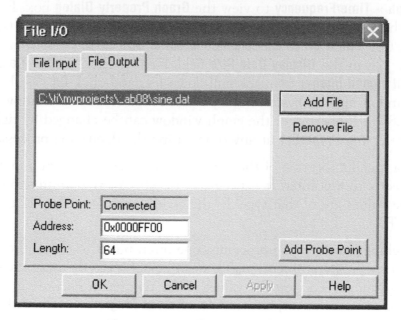

Figure L8-8: File I/O window.

Finally, enter the memory location in the **Address** field and the data type in the **Length** field. For storing the data in short format, 64 words need to be stated in the length field for 128 short data. A probe point connected to a PC file is shown in Figure L8-8. The probe point capability can be used to simulate the execution of a program in the absence of a live signal. A valid PC file should have the correct file header and extension. The file header should conform to the following format:

```
MagicNumber Format StartingAddress PageNum Length
```

MagicNumber is fixed at 1651. Format indicates the format of samples in the file: 1 for hexadecimal, 2 for integer, 3 for long, and 4 for float. StartingAddress and PageNum are determined by CCS when a stream of data is saved into a PC file. Length indicates the number of samples in memory. A valid data file should have the extension *.dat*. Data files having the same format can be transferred by choosing **File » Data » Load**... instead. However, data transfer with this capability of CCS needs to be activated manually, whereas the probe point does data updates automatically.

A graphical display of data often provides better feedback about the behavior of a program. CCS provides a signal analysis interface to monitor a signal. Let us display the array of values at 0x0000FF00 as a signal or a time graph. To do so, select **View » Graph » Time/Frequency** to view the **Graph Property Dialog** box. Field names appear in the left column. Go to the **Start Address** field, click it, and type 0x0000FF00. Then, go to the **Acquisition Buffer Size** field, click it, and enter 128. Also, enter 128 in the **Display Data Size** field. Finally, click on **DSP Data Type**, select **16-bit signed integer** from the pull-down list, and click **OK** (see Figure L8-9). A graph window appears based on the properties selected. This window is illustrated in Figure L8-10. Properties of the graph window can be changed by right-clicking on it and selecting **Properties** at any time during the debugging process.

To access a specific location of the DSP memory, one can assign a memory address directly to a pointer. It is necessary to typecast the pointer to short because the values are of that type. In the code shown, a pointer is assigned to 0x0000FF00.

When developing and testing programs, one often needs to check the value of a variable during program execution. This can be achieved by using breakpoints and watch windows. To view the values of the pointer in *sinewave.c* before and after the pointer assignment, choose **File » Reload Program** to reload the program.

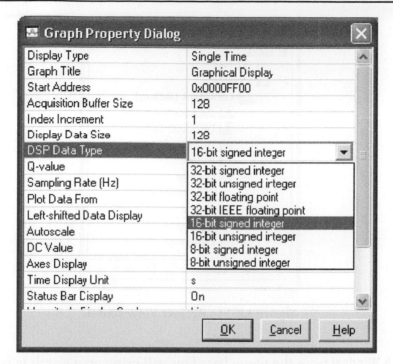

Figure L8-9: Graph Property Dialog box.

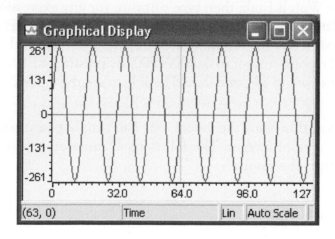

Figure L8-10: Graphical Display window.

Then, double-click on *sinewave.c* in the **Project View** panel to bring up the source file, as shown in Figure L8-11. You may wish to make the window larger to see more of the file in one place. Next, put the cursor on the line that reads output = (short *) 0x0000FF00 and press <F9> to set a breakpoint. To open a watch window, choose

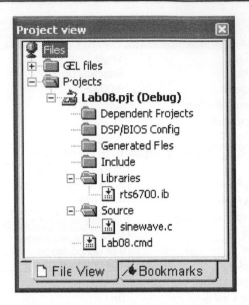

Figure L8-11: Project View panel.

View » Watch Window from the menu bar. This will bring up a watch window with the local variables listed in the Watch Locals tab. To add a new expression to the watch window, select the Watch 1 tab; then type `output` (or any expression you desire to examine) in the **Name** column. Then, choose **Debug » Run** or press <F5>. The program stops at the breakpoint and the watch window displays the value of the pointer. This is the value before the pointer is set to 0x0000FF00. By pressing <F10> to step over the line, or the shortcut button ⬛, one should be able to see the value 0x0000FF00 in the watch window.

To add a C function that sums the values, one can simply pass a pointer to an array and have a return type of integer. The following C function can be used to sum the values and return the result:

```
#include <stdio.h>
#include <math.h>

#define PI        3.141592

void main()
{
      short i, gain;
      int ret;
```

```
        float fs, f;
        float y[3], a[2], b1, x;

        short *output;
        output = (short *) 0x0000FF00;

        // Coefficient Initialization
        fs = 8000;                          // Sampling frequency
        f = 500;                            // Signal frequency
        gain = 100;                         // Amplitude gain

        a[0] = -1;
        a[1] = 2 * cos(2*PI*f/fs);
        b1 = gain;

        // Initial Conditions

        y[1] = y[2] = 0;
        x = 1;

        printf("BEGIN\n");

        for (i=0; i<128; i++)
        {
                y[0] = b1*x + a[1]*y[1] + a[0]*y[2];
                y[2] = y[1];
                y[1] = y[0];
                x = 0;

                output[i] = (short) y[0];
        }
        ret = sum(output,128);

        printf("Sum = %d\n", ret);
        printf("END\n");
}
int sum(const short* array,int N)
{
        int count,sum;
        sum = 0;

        for(count=0 ; count<N ; count++)
                sum += array[count];

        return(sum);
}
```

As part of the debugging process, it is normally required to benchmark or time a program. In this lab, let us determine how much time it takes for the function sum() to run. To achieve this benchmarking, reload the program and choose **Profile » Setup**. This will bring up the **Profile Setup** panel. Click ⏱ at the top of the panel to enable profiling. Then, click 🖉 to enable profiling of all the functions. To view and modify the profiling function list, switch to the **Ranges** tab. Remove the unnecessary function main() from the list by highlighting the function name under **Function » Enabled** and press the spacebar. Notice that the function main() now appears in gray under **Function » Disabled**. Finally, press <F5> to run the program and choose **Profile » Viewer** to examine the number of cycles for sum(), as shown in Figure L8-12 (the exact number may vary slightly from the one shown). This is the number of cycles it takes to execute the function sum() .

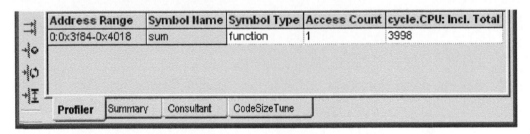

Address Range	Symbol Name	Symbol Type	Access Count	cycle.CPU: Incl. Total
0:0x3f84-0x4018	sum	function	1	3998

Profiler Summary Consultant CodeSizeTune

Figure L8-12: Profile window.

There is another way to benchmark codes using breakpoints. Double-click on the file sinewave.c in the **Project View** panel and choose **View » Mixed Source/ASM** to list the assembled instructions corresponding to the C code lines. Set a breakpoint at the calling line by placing the cursor on the line that reads ret = sum(point,128); then press <F9> or double-click **Selection Margin** located on the left side of the editor. Set another breakpoint at the next line, as indicated in Figure L8-13. Once the breakpoints are set, choose **Profile » Clock » Enable** to enable a profiler clock. Then, choose **Profile » Clock » View**, and a **Profile Clock** icon ⏱ will appear at the lower right corner of the CCS window. Press <F5> to run the program. When the program is stopped at the first breakpoint, reset the clock by double-clicking the **Profile Clock** icon. Finally, click **Step Out** or **Run** in the **Debug** menu to execute and stop at the second breakpoint. Examine the number of clock cycles displayed on the right side of the icon. It should be close to the breakpoint approach. The difference in the number of cycles between the breakpoint and the profile approaches is originated from the extra procedures for calling functions, e.g., passing arguments to function, storing return address, branching back from function, etc.

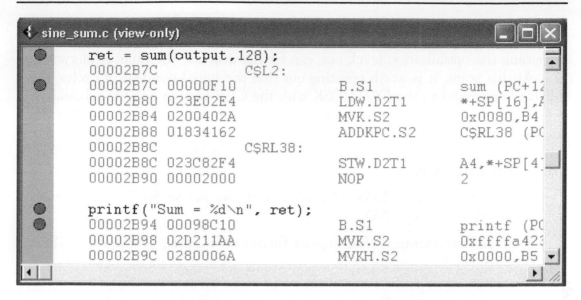

```
{ sine_sum.c (view-only)                                          [_][□][X]
   ret = sum(output,128);
   00002B7C           C$L2:
   00002B7C  00000F10                    B.S1            sum (PC+12
   00002B80  023E02E4                    LDW.D2T1        *+SP[16],A
   00002B84  0200402A                    MVK.S2          0x0080,B4
   00002B88  01834162                    ADDKPC.S2       C$RL38 (PC
   00002B8C           C$RL38:
   00002B8C  023C82F4                    STW.D2T1        A4,*+SP[4]
   00002B90  00002000                    NOP             2

   printf("Sum = %d\n", ret);
   00002B94  00098C10                    B.S1            printf (PC
   00002B98  02D211AA                    MVK.S2          0xffffa423
   00002B9C  0280006A                    MVKH.S2         0x0000,B5
```

Figure L8-13: Profiling code execution time using breakpoints.

A workspace containing breakpoints, streaming data, graphs, and watch windows can be stored and recalled later. To do so, choose **File » Workspace » Save Workspace As**. This will bring up the **Save Workspace** window. Type the workspace name in the **File name** field; then click **Save**.

Table L8-1 provides the number of cycles needed to run the sum() function using several different builds. When a program is too large to fit into the internal memory, it has to be placed into the external memory. Although the program in this lab is small enough to fit in the internal memory, it is also placed in the external memory to show the change in the number of cycles. To move the program into the external memory, open the command file *Lab08.cmd* and replace the line .text > PMEM with .text > BMEM. As seen in Table L8-1, this build slows

Table L8-1: Number of Cycles for Different Builds

Type of Build	Number of Cycles
C program in external memory	82573
C program in internal memory	3998
-o0	1838
-o1	1814
-o2	172

down the execution to 82573 cycles. In the second build, the program resides in the internal memory, and the number of cycles is hence reduced to 3998. By increasing the optimization level, one can further decrease the number of cycles to 172. At this point, it is worth pointing out that the stated numbers of cycles in this lab correspond to the C6713 DSK with the CCS version 3.0. One should realize that the numbers of cycles will vary depending on the DSK target board and CCS version used.

L8.4 Bibliography

[1] Texas Instruments, *TMS320C6000 Code Composer Studio Tutorial*, Literature ID# SPRU 301C, 2000.

[2] Texas Instruments, *Code Composer Studio User's Guide*, Literature ID# SPRU 328B, 2000.

LabVIEW DSP Integration

A DSP system designed in LabVIEW can be placed entirely or partially on a hardware platform. This chapter discusses the implementation process on a DSP hardware platform consisting of a TMS320C6713 or TMS320C6416 DSK board. Such an implementation or integration is made possible by using the LabVIEW toolkit DSP Test Integration for TI DSP.

9.1 Communication with LabVIEW: Real-Time Data Exchange (RTDX)

Communication between LabVIEW and a C6x DSK board is achieved by using the Real-Time Data Exchange (RTDX™) feature of the TMS320C6x DSP. This feature allows one to exchange data between a DSK board and a PC host (running LabVIEW) without stopping program execution on the DSP side. This data exchange is done either via the Joint Test Action Group (JTAG) connection or the Universal Serial Bus (USB) port emulating the JTAG connection.

RTDX can be configured in two modes: non-continuous and continuous. In non-continuous mode, data are written to a log file on the host. This mode is normally used for recording purposes. In continuous mode, data are buffered by the RTDX host library. This mode is normally used for continuously displaying data. Here, so that one can view the processed data on the PC/LabVIEW side, RTDX is configured to be in continuous mode.

9.2 LabVIEW DSP Test Integration Toolkit for TI DSP

The DSP Test Integration for TI DSP toolkit provides a set of VIs which enable interfacing between LabVIEW and Code Composer Studio [1]. The VIs provided in this toolkit are categorized into two groups: CCS Automation and CCS Communication. These VI groups are listed in Table 9-1.

Table 9-1: List of VIs in the LabVIEW DSP Test Integration Toolkit

CCS Automation VIs	CCS Communication VIs
CCS Open Project VI	CCS RTDX Read VI
CCS Build VI	CCS RTDX Write VI
CCS Download Code VI	CCS RTDX Enable VI
CCS Run VI	CCS RTDX Enable Channel VI
CCS Halt VI	CCS RTDX Disable VI
CCS Close Project VI	CCS RTDX Disable Channel VI
CCS Window Visibility VI	CCS Memory Read VI
CCS Reset VI	CCS Memory Write VI
	CCS Symbol to Memory Address VI

The VIs in the CCS Automation group allow automating the CCS code execution steps through LabVIEW. They include (a) open CCS, (b) build project, (c) reset CPU, (d) load program, (e) run code, (f) halt CPU, and (g) close CCS. The flow of these steps is the same as those in CCS.

The VIs in the CCS Communication group allow exchange of data through the RTDX channel. For example, the CCS RTDX write VI and CCS RTDX read VI are used for writing and reading data to and from the DSP side, respectively. Note that these VIs are polymorphic. Therefore, data types (i.e., single precision, double precision, or integer) and data formats (i.e., scalar or array) should be matched in LabVIEW and CCS in order to establish a proper LabVIEW DSP integration.

9.3 Combined Implementation: Gain Example

In this section, a LabVIEW DSP integration example is presented to show the basic steps that are required for a combined LabVIEW and DSP implementation. From the main dialog box of LabVIEW, open the NI Example Finder, shown in Figure 9-1, by choosing **Help » Find Examples**.

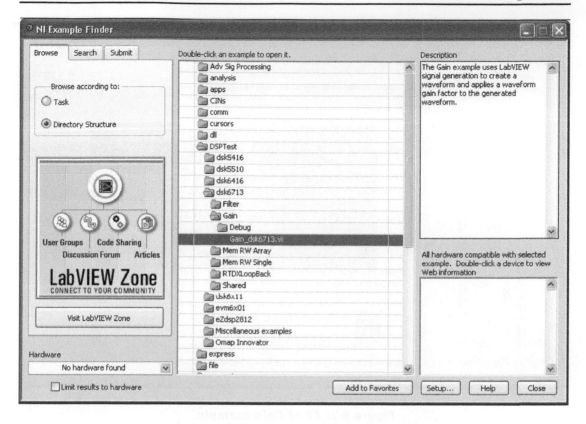

Figure 9-1: NI Example Finder—Gain example.

Open the `Gain_dsk6713` VI by clicking on **Directory Structure** from the category **Browse according to** of the **Browse** tab and by choosing **DSPTest » dsk6713 » Gain » Gain_dsk6713.vi**. If using a C6416 DSK, open the *dsk6416* folder.

In this example, an input signal and a gain factor are sent from the LabVIEW side to the DSP side. On the DSP side, the input signal is multiplied by the gain factor and then sent back to the LabVIEW side. The gain factor, the frequency of the input signal, and the signal type can be altered by the controls specified in the FP. Also, the original and scaled signals are displayed in the FP, as shown in Figure 9-2.

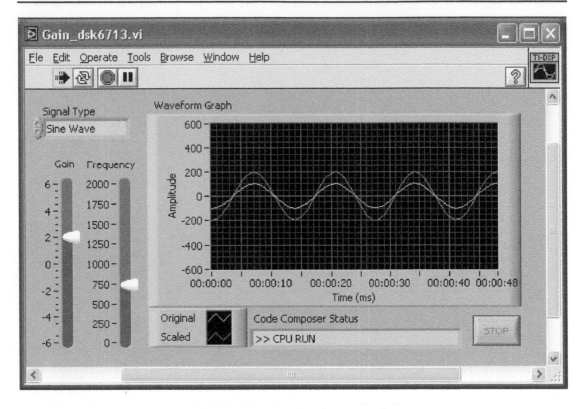

Figure 9-2: FP of Gain example.

9.3.1 LabVIEW Configuration

To better understand the LabVIEW DSP integration process, let us examine the BD of the Gain_dsk6713 VI, which is shown in Figure 9-3.

Two major sections are associated with this BD. The first section consisting of a Stacked Sequence structure, shown to the left side of the While Loop, corresponds to the CCS automation process. This section includes a CCS Open Project VI, a CCS Build VI, a CCS Reset VI, a CCS Download Code VI, and a CCS Run VI. In addition, a CCS Halt VI and a CCS Close Project VI, shown to the right side of the While Loop, are a part of the CCS automation process. The three functions (Strip Path, Build Path, and Current VI's Path) of the File I/O palette (**Functions » Programming » File I/O**) are used in the Stacked Sequence structure to create a file path to a CCS project file that can be opened from the CCS side. With these VIs and functions in place, the process of opening CCS, building a project, loading a program to the DSP, and running it on

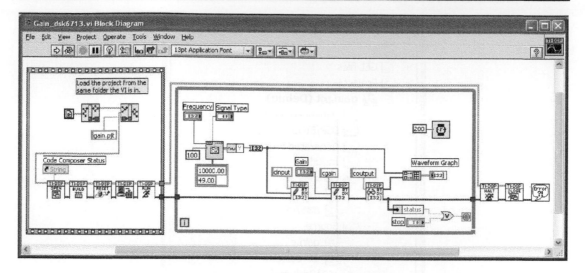

Figure 9-3: BD of Gain example.

the DSP can be controlled from the LabVIEW side. The CCS Automation VIs are located in the DSP Test Integration Palette (**Functions » Addons » DSP Test Integration**). Note that the CCS automation process just described can be used for all the LabVIEW DSP integration examples presented in Lab 9.

The second section of the BD shown in the While Loop involves signal generation and CCS RTDX communication. The Basic Function Generator VI (**Functions » Signal Processing » Waveform Generation**) is used to generate waveform samples. Two CCS RTDX read VIs and one CCS RTDX write VI are located in the While Loop. The channel name of each CCS RTDX VI is wired to this VI. This allows the generated samples to be continuously sent to the DSP side, and the DSP processed samples to be continuously read from the DSP side. The scaled signal is displayed in a waveform graph. Note that the original and scaled signals are of array type, while the gain factor is scalar. Thus, one of the CCS RTDX read/write VIs is set to 32-bit integer array, indicated by [I32] on its icon, and the other is set to 32-bit integer, indicated by I32 on its icon.

9.3.2 DSP Configuration

A CCS project implemented on the DSP side should include four components: a linker command file, an interrupt service table which defines the interrupt vector for RTDX, the RTDX library along with the run-time support library, and the source code, as shown in Figure 9-4.

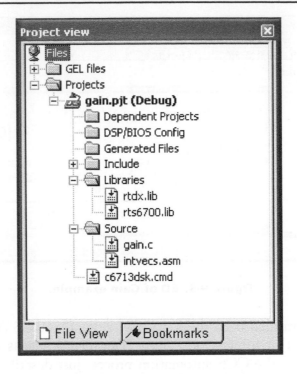

Figure 9-4: Project view of CCS.

The C source code of the Gain example running on the DSP side is as follows:

```c
#include <rtdx.h>                       /* RTDX                   */
#include "target.h"                     /* TARGET_INITIALIZE()    */

#define kBUFFER_SIZE 49

RTDX_CreateInputChannel(cinput);
RTDX_CreateInputChannel(cgain);
RTDX_CreateOutputChannel(coutput);

// Gain value scales the waveform
void Gain (int *output, int *input, int gain)
{
      int i;
      for(i=0; i<kBUFFER_SIZE; i++)
            output[i]=input[i]*gain;
}
void main()
{
      int input[kBUFFER_SIZE];
```

```
int output [kBUFFER_SIZE];
int gain = 10;

// Target initialization for RTDX
TARGET_INITIALIZE();

/* enable RTDX channels */
RTDX_enableInput(&cgain);
RTDX_enableInput(&cinput);
RTDX_enableOutput(&coutput);

for (;;) /* Infinite message loop. */
{
        /* Read new gain value if one exists */
        if (!RTDX_channelBusy(&cgain))
                RTDX_readNB(&cgain, &gain, sizeof(gain));
        /* Wait for input waveform */
        while(!RTDX_read(&cinput, input, sizeof(input)));

        Gain (output, input, gain);

        /* Write scaled data back to host. */
        RTDX_write(&coutput, output, sizeof(output));
}
}
```

In this code, several application programming interfaces (APIs), which are part of the CCS RTDX library, are used to allow data exchange between the DSP and LabVIEW side. First, the RTDX_CreateInputChannel() and RTDX_CreateOutputChannel() APIs are used to declare the input and output channels. Second, the DSP board is initialized with the TARGET_INITIALIZE() API. Both of the RTDX channels are enabled by the RTDX_enableInput() and RTDX_enableOutput() APIs. To get scalar data from the LabVIEW to the DSP side, one needs to use the RTDX_readNB() API, with the arguments being channel, buffer pointer, and buffer size. In addition, one needs to use the RTDX_read() API, with the arguments being channel, array pointer, and array size. The RTDX_write() API is used to send data back to the LabVIEW side.

Bear in mind that the name assigned to the RTDX communication channel should be the same as the one used in LabVIEW. Also, the data types of polymorphic VIs, i.e., CCS RTDX Read VI and CCS RTDX Write VI, as well as the array lengths should be the same as the ones defined in LabVIEW. For example, the input array input[ ] in the preceding source code should be defined as follows:

```
int input[kBUFFER_SIZE];
```

That is, input[ ] must be declared as a 32-bit integer, and the array size must be configured to be kBUFFER_SIZE, which is specified as 49 at the beginning of the Gain sample code.

9.4 Bibliography

[1] National Instruments, *LabVIEW DSP Test Toolkit for TI DSP User's Manual*, Literature Number 323452A-01, 2002.

Lab 9: DSP Integration Examples

This lab includes four DSP integration examples. These examples correspond to the DSP systems built by LabVIEW in the previous labs, i.e., digital filtering, integer arithmetic, adaptive filtering, and frequency processing.

L9.1 CCS Automation

Figure L9-1 illustrates the CCS automation process. In this lab, all the examples are assumed to have the sub-diagrams shown to the left and right of the While Loop and thus are not explicitly mentioned.

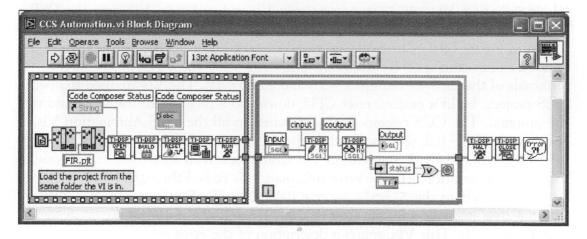

Figure L9-1: Generic structure of CCS Automation.

Let us explain the CCS automation process in more detail. In order to specify a project to be used by the CCS Automation VIs, one should build a file path to the project file. Two methods of creating a path are mentioned here. The first method involves using a relative path by assuming that the CCS project file is located in the folder that the VI resides. Place a Current VI's Path function (**Functions » Programming » File I/O » File Constants » Current VI's Path**) in the BD to get the entire file path of the project file and wire the output of this VI to the path terminal of a Strip Path function (**Functions » Programming » File I/O » Strip Path**), which returns a stripped path by removing the VI's name from the path. The stripped path is wired to the base path terminal of the Build Path function (**Functions » Programming » File I/O » Build Path**). This function appends the name of the file, wired to the name or relative path terminal of the function as a string constant, to the stripped path. Now, the output of the function indicates the entire path of the CCS project file. The created file path is then wired to the `Path to Project` terminal of the `CCS Open Project` VI to allow access by the CCS Automation VIs.

The second method of creating a project path involves using an absolute path. Wire a `Path Constant` (**Functions » Programming » File I/O » File Constants » Path Constant**) to the `Path to Project` terminal of the `CCS Open Project` VI or create a file constant by right-clicking and choosing **Create » Constant** on the `Path to Project` terminal of the VI. Enter the absolute path of the CCS project file in the `Path Constant`. The absolute path can also be generated by browsing the project file path. To do this, right-click on the `Path Constant` and choose **Browse for Path**... from the shortcut menu. A file dialog box appears to select the path via file browsing.

Next, place the CCS Automation VIs (`CCS Open Project` VI, `CCS Build` VI, `CCS Reset` VI, `CCS Download Code` VI, and `CCS Run` VI) from the DSP Test Integration Palette (**Functions » Addons » DSP Test Integration**) in the order shown in Figure L9-1. These VIs are wired to each other via the terminals of the `CCS references out` (or `dup CCS references`) and `error out` to the terminals of the `CCS references in` and `error in`. The VIs are used to open a CCS project, build a project, reset CPU, download a program to the DSP, and run the program. The CCS references cluster, wired to all the CCS Automation VIs, contains the CCS IDE references, while the `error in/out` cluster carries the error information of the CCS Automation VIs. Consequently, if an error occurs in one of the CCS Automation VIs, the error information is passed through the CCS Automation VIs to the `Simple Error Handler` VI (**Functions » Programming » Dialog & User Interface » Simple Error Handler**) located at the end of the CCS Automation VIs. This VI displays a description of the error.

A `String Indicator` is placed in the FP in order to display the current status of the CCS automation process. A `Control Reference` for this indicator is created by right-clicking on it and choosing **Create » Reference** from the shortcut menu. This reference should be wired to the `Status String Refnum` terminal of the `CCS Open Project` VI in order to post a status string to this indicator.

Now, let us explain the exchange of data between LabVIEW and the DSP. Data are continuously exchanged using the CCS Communication VIs, `CCS RTDX Read` VI and `CCS RTDX Write` VI, when both the VI and CCS are running. As mentioned earlier, data types should be carefully configured on the LabVIEW and CCS sides, since the `CCS RTDX Read` VI and the `CCS RTDX Write` VI are data type polymorphic. The read/write data type can be specified from the **Select Type** menu as part of the shortcut menu. This menu can be brought up by right-clicking on the `CCS`

RTDX Read VI or CCS RTDX Write VI. Another way to change the data type is to use a Polymorphic VI Selector. This selector can be displayed by right-clicking on it and choosing **Visible Items » Polymorphic VI Selector** from the shortcut menu. The string constants indicate the names of the RTDX channels that are wired to the CCS RTDX Read VI or CCS RTDX Write VI.

The execution of the While Loop structure is stopped by pressing a stop button in the FP or if an error is generated by the CCS Automation or CCS Communication VIs. In such cases, the CCS needs to be halted and closed. This is done by locating and wiring a CCS Halt VI to a CCS Close Project VI. These VIs appear to the right side of the While Loop structure.

L9.2 Digital Filtering

In this section, the filtering code written in C is used to run the filtering block or component of the Lab 4 filtering system on the DSP.

L9.2.1 FIR Filter

The BD of the FIR lowpass filtering system that was built in Lab 4 is illustrated in Figure L9-2.

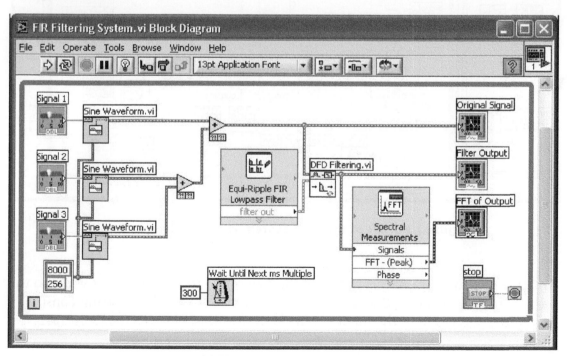

Figure L9-2: Filtering system in Lab 4.

Let us modify this BD to send the generated samples to the DSP and then to receive the filtered samples from the DSP. This is achieved by inserting the CCS automation process to the left and right sections of the While Loop structure. As indicated in Figure L9-3, a portion of the DFD Filter VI is replaced with the CCS RTDX Write VI and the CCS RTDX Read VI. Both of these VIs are configured to write and read single-precision floating-point array data, which means configuring the polymorphic VIs as CCS RTDX Write Array SGL and CCS RTDX Read Array SGL; refer to Figure L9-3. Consider that the number of samples in the sampling info cluster is reduced to 128 in order to reduce the time associated with the RTDX communication.

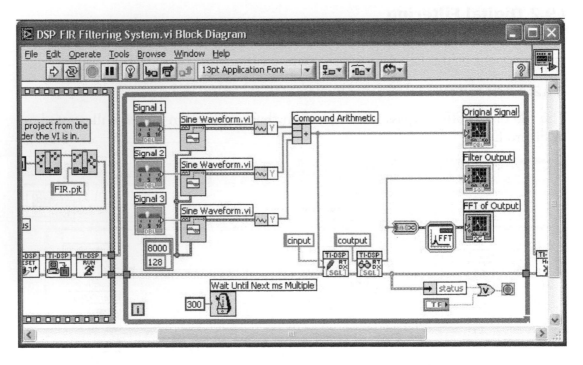

Figure L9-3: BD of filtering system with DSP integration.

An array of signal samples consisting of the sum of the three sinusoids is wired to the Data terminal of the CCS RTDX Write Array SGL VI. Also, a string constant containing the name of the input channel, cinput, is wired to the Channel terminal.

In the CCS RTDX Read Array SGL VI, the data transmitted via RTDX is read from the Data terminal of the VI. This terminal is wired to a waveform graph as well as to a Spectral Measurements Express VI for frequency analysis.

To check the status of errors generated by the CCS Automation or CCS Communication VIs, observe the status element of the error cluster. This is made possible by locating an Unbundle By Name function (**Functions » Programming » Cluster & Variant » Unbundle By Name**). Wire the error out cluster from the CCS RTDX Read Array SGL VI to the Unbundle By Name function. This way, the status element of the error out cluster is selected by default. The result of an OR operation of two Boolean values, corresponding to the status element of the cluster and a stop button, is wired to the conditional terminal of the While Loop. Whenever the stop button is pressed or an error occurs while accessing CCS or communicating via RTDX, the execution of the loop stops.

Notice the importance of the timeout value of the CCS RTDX Read Array SGL VI. If the RTDX communication speed is too slow or the number of data samples is large, the timeout value should be changed to avoid getting an RTDX error, as shown in Figure L9-4. The default timeout value is 2000. To change the timeout value, wire a Numeric Constant to the timeout terminal of the CCS RTDX Read Array SGL VI and enter a desired timeout value in milliseconds. Save the completed VI as *DSP FIR Filtering System.vi*.

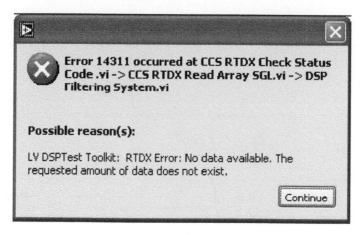

Figure L9-4: RTDX error.

Next, let us discuss how to create a project in CCS. Create a new project and name it *FIR.pjt*. Add the linker command file *c6713dsk.cmd*, the interrupt service vector *intvecs.asm*, the source code *FIR.c*, and the library files *rtdx.lib (CCStudio\C6000\ rtdx\lib)* and *rts6700.lib (CCStudio\C6000\cgtools\lib)* into the project. The linker command file and interrupt service vector are located in the folder *DSK6713\ Shared*. This folder also includes the header file *target.h*. The path to this folder needs to be added in **Include Search Path** of **Build Options**, as shown in Figure L9-5.

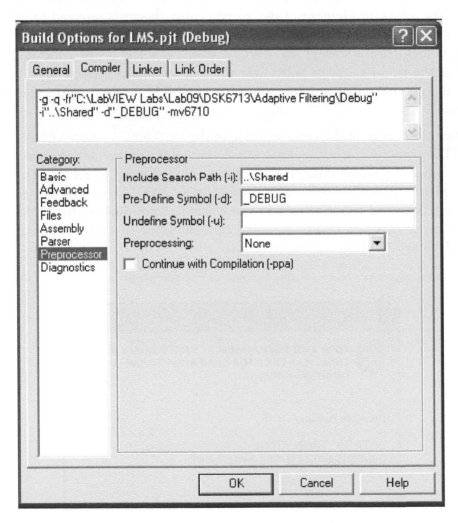

Figure L9-5: Build Options of CCS.

It is worth mentioning that, if a C6416 DSK is used, the files and libraries to be added to the project are different from those when using a C6713 DSK. The corresponding path of the linker command (*c6416dsk.cmd*) and interrupt service vector (*intvecs6416.asm*) files for the DSK6416 board is *DSK6416\Shared*. The libraries that need to be added to the project are *rtdx64xx.lib (CCStudio\C6000\ rtdx\lib)* and *rts6400.lib (CCStudio\C6000\cgtools\lib)*.

The FIR filtering C source code is as follows:

```c
#include <rtdx.h>              /* RTDX               */
#include "target.h"            /* TARGET_INITIALIZE() */

#define kBUFFER_SIZE 128
#define N 15

float b[N] = {-0.008773, 0.0246851, 0.0217041, -0.0396942,
    -0.0734726, 0.0560876, 0.305969, 0.437322, 0.305969,
     0.0560876, -0.0734726, -0.0396942, 0.0217041, 0.0246851,
    -0.008773};
float samples[N];

RTDX_CreateInputChannel(cinput);
RTDX_CreateOutputChannel(coutput);

void FIR(float *input, float *output)
{
    int i, j;
    float result;

    for(j=0; j<kBUFFER_SIZE; j++)
    {
        for(i=N-1; i>0; i--)
            samples[i] = samples[i-1];

        samples[0] = input[j];

        result = 0;
        for(i=0; i<N ; i++)
            result += samples[i] * b[i];

        output[j] = result;
    }
}
```

```
void main()
{
        float input[kBUFFER_SIZE];
        float output[kBUFFER_SIZE];
        int i;

        for(i=0; i<N; i++)
                samples[i] =0;

        // Target initialization for RTDX
        TARGET_INITIALIZE();

        /*enable RTDX channels */
        RTDX_enableInput(&cinput);
        RTDX_enableOutput(&coutput);

        for (;;) /* Infinite message loop. */
        {
                while(!RTDX_read(&cinput, input, sizeof(input)));

                FIR(input, output);

                /* Write filtered data back to host. */
                RTDX_write(&coutput, output, sizeof(output));
        }
}
```

After creating the VI for the signal source and the CCS project for the FIR filtering block, run the VI from LabVIEW. One should see the outcome depicted in Figure L9-6. Notice that the amplitudes of the frequency components in the stopband (2200-4000 Hz) appear attenuated by 30 dB. This agrees with the filter specification.

L9.2.2 IIR Filter

The bandpass IIR filter designed in Lab 4 is modified here. The DSP FIR Filtering System VI used previously is modified in order to run the IIR filtering project, i.e., *IIR.pjt*, on the DSP. The modified VI is then saved as *DSP IIR Filtering System.vi*.

As mentioned in Chapter 4, by default, the DFD Classical Filter Design Express VI of the DFD toolkit provides the IIR filter coefficients in the second-order cascade form. In the following C source code, the IIR filter comprises

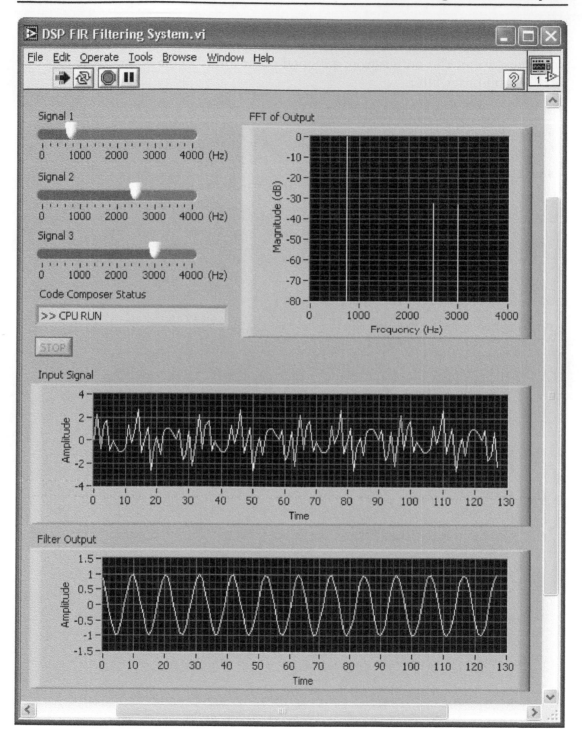

Figure L9-6: FP of FIR filtering system with DSP integration.

three second-order IIR filters in cascade. The advantage of the second-order cascade form lies in its lower sensitivity to coefficient quantization. In this implementation, the output from a second-order filter becomes the input to a next second-order filter.

```
#include <rtdx.h>                        /* RTDX                */
#include "target.h"                      /* TARGET_INITIALIZE() */

#define kBUFFER_SIZE 128

float a1[2]={-0.955505, 0.834882};
float b1[3]={0.545337, -0.735242, 0.545337};

float a2[2]={0.954255, 0.834810};
float b2[3]={0.545337, 0.734702, 0.545337};

float a3[2]={-0.000622, 0.372609};
float b3[3]={0.545337, 0, -0.545337};

float IIRwindow1[3] = {0,0,0};
float y_prev1[2] = {0,0};

float IIRwindow2[3] = {0,0,0};
float y_prev2[2] = {0,0};

float IIRwindow3[3] = {0,0,0};
float y_prev3[2] = {0,0};

RTDX_CreateInputChannel(cinput);
RTDX_CreateOutputChannel(coutput);

void main()
{
        float input[kBUFFER_SIZE];
        float output[kBUFFER_SIZE];
        int i, n;
        float ASUM, BSUM;

        // Target initialization for RTDX
        TARGET_INITIALIZE();

        /*enable RTDX channels */
        RTDX_enableInput(&cinput);
        RTDX_enableOutput(&coutput);
```

```
for (;;)          /* Infinite message loop. */
{
        while(!RTDX_read(&cinput, input, sizeof(input)));

        // IIR filtering

        for(i=0; i<kBUFFER_SIZE; i++)
        {
                // Stage #1

                for(n=2; n>0; n--)
                    IIRwindow1[n] = IIRwindow1[n-1];

                IIRwindow1[0] = input[i];

                BSUM = 0;
                for(n=0; n<=2; n++)
                {
                        BSUM += b1[n] * IIRwindow1[n];
                }
                ASUM = 0;
                for(n=0; n<=1; n++)
                {
                    ASUM += a1[n] * y_prev1[n];
                }

                y_prev1[1] = y_prev1[0];
                y_prev1[0] = BSUM - ASUM;

                // Stage #2

                for(n=2; n>0; n--)
                    IIRwindow2[n] = IIRwindow2[n-1];

                IIRwindow2[0] = y_prev1[0];

                BSUM = 0;
                for(n=0; n<=2; n++)
                {
                        BSUM += b2[n] * IIRwindow2[n];
                }
                ASUM = 0;
                for(n=0; n<=1; n++)
                {
                    ASUM += a2[n] * y_prev2[n];
                }
```

```
                    y_prev2[1] = y_prev2[0];
                    y_prev2[0] = BSUM - ASUM;


                    // Stage #3

                    for(n=2; n>0; n--)
                        IIRwindow3[n] = IIRwindow3[n-1];

                    IIRwindow3[0] = y_prev2[0];

                    BSUM = 0;
                    for(n=0; n<=2; n++)
                    {
                            BSUM += b3[n] * IIRwindow3[n];
                    }
                    ASUM = 0;
                    for(n=0; n<=1; n++)
                    {
                        ASUM += a3[n] * y_prev3[n];
                    }
                        output[i] = BSUM - ASUM;
                    y_prev3[1] = y_prev3[0];
                    y_prev3[0] = output[i];
            }

            /* Write data back to host. */
            RTDX_write(&coutput, output, sizeof(output));
    }
}
```

The output of the IIR bandpass filter is depicted in Figure L9-7. Considering that the passband of this filter is between 1333 and 2667 Hz, the amplitude of any signal in the stopband is attenuated by about 25 dB, which matches the outcome in Lab 4.

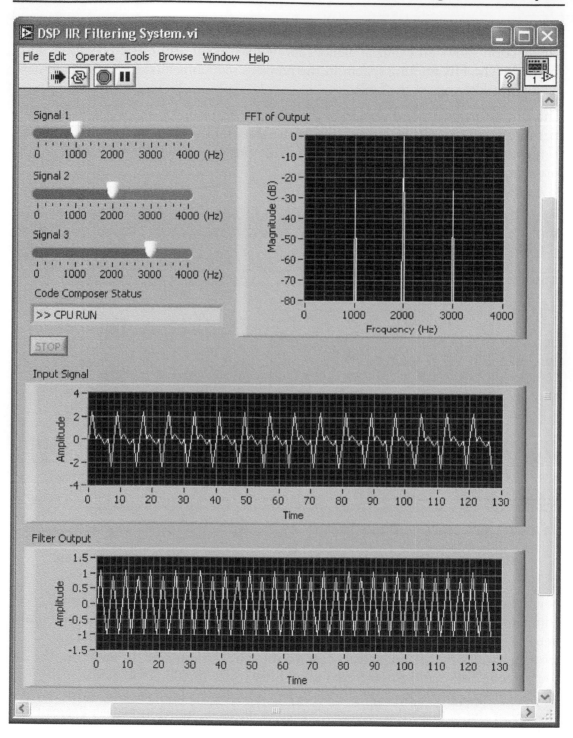

Figure L9-7: FP of IIR filtering system with DSP integration.

L9.3 Fixed-Point Implementation

This section shows an example demonstrating fixed-point arithmetic operations on the DSP. The FIR filtering system in the previous section is modified here to achieve fixed-point filtering on the DSP.

The BD of the fixed-point version of the FIR filtering system is shown in Figure L9-8. In this BD, the amplitude of the sum of the three sinusoids is multiplied by 10000 to represent its value as a 16-bit integer while not exceeding the representable range of 16-bit integer numbers.

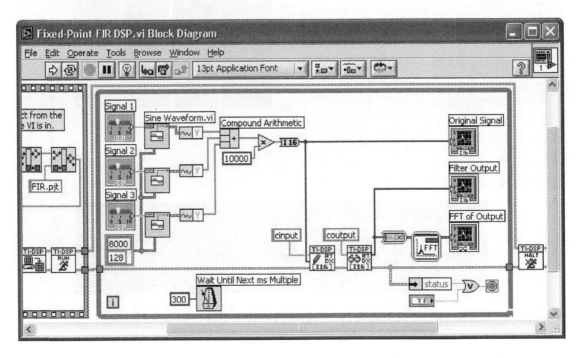

Figure L9-8: BD of fixed-point FIR filtering system with DSP integration.

For the simulator case, use *rtdxsim.lib* (*CCStudio\C6000\rtdx\lib*). The linker command and the interrupt service vector files, *c6416dsk.cmd* and *intvecs6416.asm*, need to be included from the *Simulator/Shared* folder for the C6416 simulator, likewise *c6713dsk.cmd* and *intvecs.asm* for the C6713 simulator. In addition, the source code *FIR.c* and the corresponding library file *rts6400.lib* or *rts6700.lib* (*CCStudio\ C6000\cgtools\lib*) need to be added to the project accordingly. Note that the path to the *Simulator/Shared* folder is included in **Include Search Path** of **Build Options**.

The source code of the fixed-point version of the FIR filtering system is shown here. In this code, the filter coefficients originally expressed in floating-point format are first converted into Q15 format. Then, they are scaled by one-half to avoid overflows. The number of scaling is determined to be one by considering that all the inputs are 1's, as discussed in Lab 5.

```c
#include <rtdx.h>               /* RTDX                */
#include "target.h"             /* TARGET_INITIALIZE() */

#define kBUFFER_SIZE 128
#define N 15

float b[N] = {-0.008773, 0.0246851, 0.0217041, -0.0396942,
    -0.0734726, 0.0560876, 0.305969, 0.437322, 0.305969,
     0.0560876, -0.0734726, -0.0396942, 0.0217041, 0.0246851,
    -0.008773};

short samples[N];
short coeff[N];

RTDX_CreateInputChannel(cinput);
RTDX_CreateOutputChannel(coutput);

void FIR(short *input, short *output)
{
      int i, j;
      int result;

      for(j=0; j<kBUFFER_SIZE; j++)
      {
            for(i=N-1; i>0; i--)
                  samples[i] = samples[i-1];

            samples[0] = input[j];

            result = 0;
            for(i=0; i<N; i++)
                  result += ( samples[i] * coeff[i] ) << 1;

            result = result >> 16;

            // Scale the Output to Compensate Scaling of Coefficients.
            output[j] = (short) ( result << 1 );
      }
}
```

```
void main()
{
        short input[kBUFFER_SIZE];
        short output[kBUFFER_SIZE];
        int i;

        for(i=0; i<N ; i++)
                samples[i] =0;

        // Convert to Q-15
        for(i=0; i<N ; i++)
                coeff[i] = b[i] * 0x7fff;

        // Scale by Half
        for(i=0; i<N ; i++)
                coeff[i] = coeff[i] >> 1;

        // Target initialization for RTDX
        TARGET_INITIALIZE();

        /*enable RTDX channels */
        RTDX_enableInput(&cinput);
        RTDX_enableOutput(&coutput);

        for (;;)        /* Infinite message loop. */
        {
                while(!RTDX_read(&cinput, input, sizeof(input)));

                FIR(input, output);

                /* Write filtered data back to host. */
                RTDX_write(&coutput, output, sizeof(output));
        }
}
```

The multiplication of two Q15 numbers results in a Q30 format number with an extended sign bit being at the most significant bit. The extended sign bit is removed by left-shifting this output number by one bit, which makes it a Q31 format number. To store it in Q15 format, one needs to right-shift it by 16 bits.

The FP corresponding to the fixed-point DSP integration is shown in Figure L9-9. As one can see from this figure, the displays match those in the floating-point version shown in Figure L9-6.

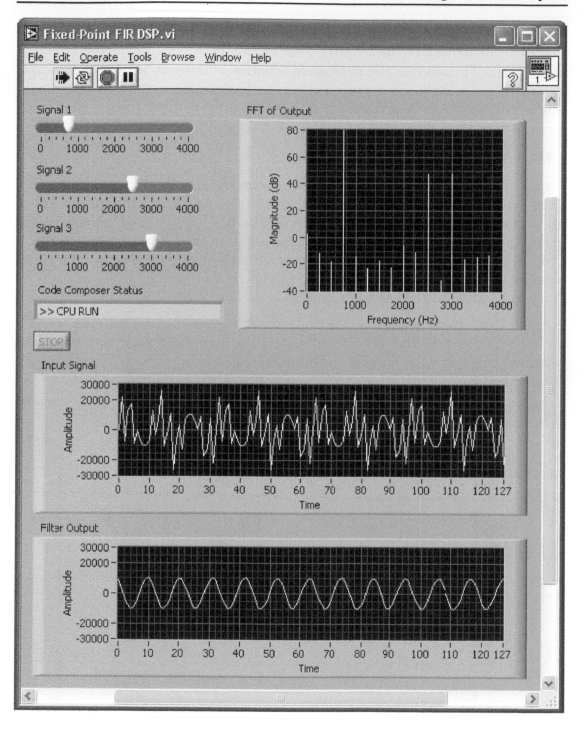

Figure L9-9: FP of fixed-point FIR filtering system with DSP integration.

L9.4 Adaptive Filtering Systems

The DSP integration of the adaptive filtering systems in Lab 6 is presented in this section. Though one can implement adaptive filtering by sending one sample at a time to the DSP, this approach is very inefficient due to the overhead associated with the RTDX communication. It is thus more efficient to send an array of input data to the DSP where point-by-point processing is performed.

L9.4.1 System Identification

An IIR filter is used to act as the unknown system by using the Butterworth Filter VI. Note that unlike the Butterworth Filter PtByPt VI used in Lab 6, this VI processes an array input. A 64-sample sinusoidal signal is used as the reference input via the RTDX channel cin1, and the output of the IIR filter is sent to the DSP via the RTDX channel cin2. The output of the LMS FIR filter and the error between the filter output and the desired output are read via the cout1 and cout2 channels, respectively.

A True Constant is wired to the init/cont terminal of the Butterworth Filter VI. This disables the initialization of the internal state of the filter, thus avoiding the group delay effect at the beginning of each output array. The BD of the system identification system with DSP integration is shown in Figure L9-10.

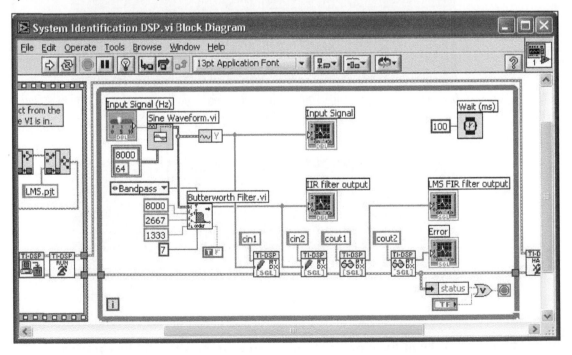

Figure L9-10: System identification with DSP integration.

The C code for performing adaptive filtering on the C6x DSP is shown here. This code updates two arrays, consisting of the coefficients and input samples, at each iteration, similar to the LabVIEW implementation.

```c
#include "target.h"
#include <rtdx.h>

#define N 32 //filter length
#define kBUFFER_SIZE 64

float h[N] = {0.0, 0.0, 0.0, 0.0, 0.0, 0.0, 0.0, 0.0, 0.0, 0.0, 0.0, 0.0,
0.0, 0.0, 0.0, 0.0, 0.0, 0.0, 0.0, 0.0, 0.0, 0.0, 0.0, 0.0, 0.0, 0.0, 0.0,
0.0, 0.0, 0.0, 0.0, 0.0};
float samples[N];

RTDX_CreateInputChannel(cin1);
RTDX_CreateInputChannel(cin2);
RTDX_CreateOutputChannel(cout1);
RTDX_CreateOutputChannel(cout2);

void main()
{
      float input1[kBUFFER_SIZE];
      float input2[kBUFFER_SIZE];
      float output[kBUFFER_SIZE];
      float e[kBUFFER_SIZE];

      int i, j;
      float stemp, stemp2;

      for(i=0; i<N ; i++)
            samples[i] = 0;

      // Target initialization for RTDX
      TARGET_INITIALIZE();

      /*enable RTDX channels */

      RTDX_enableInput(&cin1);
      RTDX_enableInput(&cin2);
      RTDX_enableOutput(&cout1);
      RTDX_enableOutput(&cout2);

      for (;;)          /* Infinite message loop. */
      {
            /* Wait for input sample */
            while(!RTDX_read(&cin1, input1, sizeof(input1)));
```

```
        while(!RTDX_read(&cin2, input2, sizeof(input2)));

        for (j=0; j<kBUFFER_SIZE; j++)
        {
                // Update array samples
                for(i=N-1; i>0; i--)
                        samples[i] = samples[i-1];

                samples[0] = input1[j];

                stemp =0;

                // FIR Filtering
                for(i=0; i<N ; i++)
                        stemp += (samples[i] * h[i]);
                output[j] = stemp;

                e[j] = input2[j] - stemp;

                stemp = (0.001 * e[j]);

                // Update Coefficient
                for(i=0; i<N; i++)
                {
                        stemp2 = (stemp * samples[i]);
                        h[i] = h[i] + stemp2;
                }
        }

        /* Write scaled data back to host. */
        RTDX_write(&cout1, output, sizeof(output));
        RTDX_write(&cout2, e, sizeof(e));
    }
}
```

The output of the IIR filter and the adaptive FIR filter are shown in Figure L9-11. The output of the adaptive FIR filter adapts to the output of the IIR filter (unknown system) when the input is changed. Notice that the speed of convergence is governed by the step size specified in the C code.

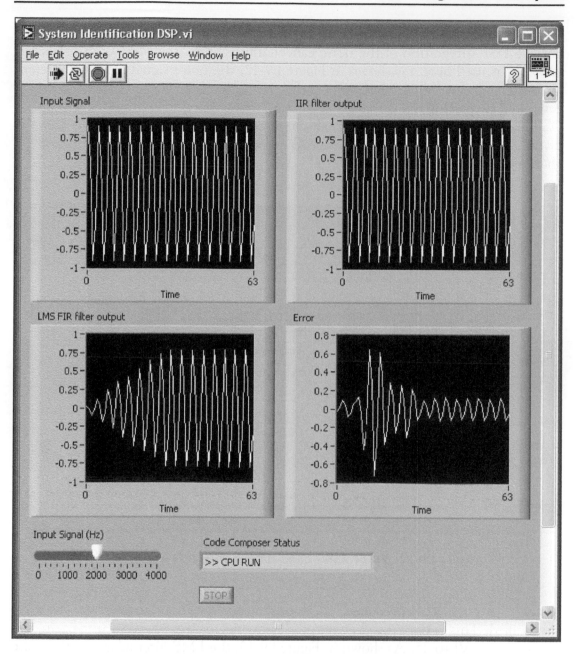

Figure L9-11: System identification with DSP integration.

L9.4.2 Noise Cancellation

For the DSP integration of the noise cancellation system, the same CCS project, *LMS.pjt*, is used here. As shown in Figure L9-12, the noise signal acts as the reference signal and is sent to the DSP via the `cin1` channel. The filtered noise signal, generated by passing the noise signal through a time-varying channel, is sent to the DSP via the `cin2` channel. The LMS filter output then becomes available from the `cout1` channel, and the noise-cancelled output signal is read from the `cout2` channel.

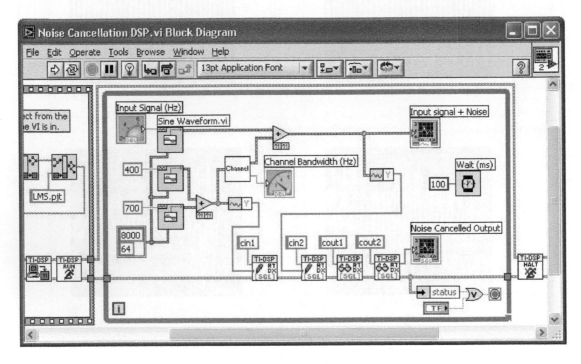

Figure L9-12: BD of noise cancellation with DSP integration.

In the `Channel` VI, introduced in Lab 6, the time duration between the steps is modified. This is done by changing the frequency of the `Basic Function Generator` VI to 25. As shown in Figure L9-13, the LMS filter adapts to the noise signal in such a way that the difference between its output and the noise corrupted signal approaches zero.

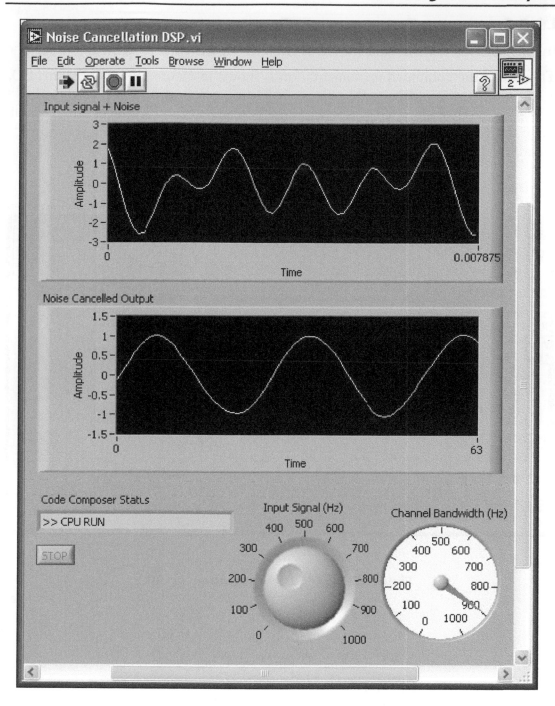

Figure L9-13: FP of noise cancellation with DSP integration.

L9.5 Frequency Processing: FFT

In this section, the DSP integration of the FFT algorithm is presented.

The BD of the combined implementation is shown in Figure L9-14(a). In this BD, a 128-sample sinusoidal signal having a 16-bit integer array format is sent to the DSP. Notice that the samples read from the DSP are in the 32-integer array format, since the FFT magnitude values are quite large, as indicated in the FP shown in Figure L9-14(b).

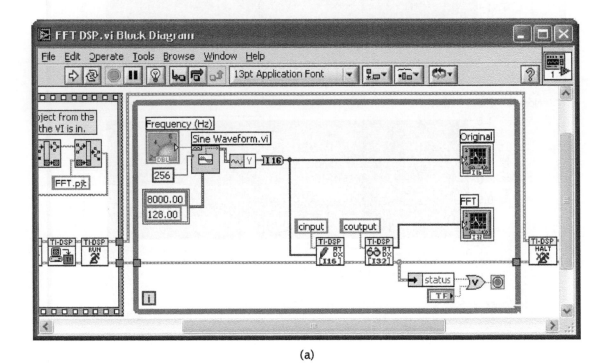

(a)

Figure L9-14: FFT DSP integration: (a) BD.

Continued

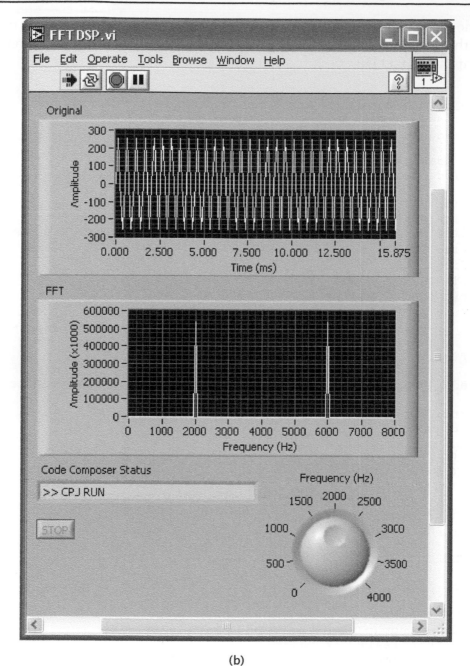

(b)

Figure L9-14 Continued: FFT DSP integration: (b) FP.

For the DSP implementation, it is required to have the C source code of the FFT algorithm presented in Chapter 7. This code, which follows, is provided on the accompanying CD [3]:

```c
#include <math.h>
#include "twiddleR.h"
#include "twiddleI.h"

#include <rtdx.h>                          /* RTDX                 */
#include "target.h"                        /* TARGET_INITIALIZE() */

#define kBUFFER_SIZE 128
#define NUMDATA       128   /* number of real data samples */
#define NUMPOINTS      64   /* number of points in the DFT, NUMDATA/2 */
#define TRUE     1
#define FALSE    0
#define BE        TRUE
#define LE        FALSE
#define ENDIAN    LE        /* selects proper endian building code
                               in Big Endian, use BE, else use LE */

#define   PI       3.141592653589793 /* definition of pi */

typedef struct {    /* define the data type for the radix-4 twiddle
                       factors */
short imag;
short real;
} COEFF;

/* BIG Endian */
#if ENDIAN == TRUE

typedef struct {
      short imag;
      short real;
} COMPLEX;

#else

/* LITTLE Endian */
typedef struct {
      short real;
      short imag;
} COMPLEX;

#endif

#pragma DATA_ALIGN(x,NUMPOINTS);
COMPLEX x[NUMPOINTS+1];                     /* array of complex DFT data */
```

```
COEFF W4[NUMPOINTS];
short g[NUMDATA];
COMPLEX A[NUMPOINTS];          /* array of complex A coefficients */
COMPLEX B[NUMPOINTS];          /* array of complex B coefficients */
COMPLEX G[2*NUMPOINTS];        /* array of complex DFT result     */
unsigned short IIndex[NUMPOINTS], JIndex[NUMPOINTS];
int count;

int magR[NUMDATA];
int magI[NUMDATA];

int output[kBUFFER_SIZE];

void make_q15(short out[], float in[], int N);
void R4DigitRevIndexTableGen(int, int *, unsigned short *,
unsigned short *);
void split1(int, COMPLEX *, COMPLEX *, COMPLEX *, COMPLEX *);
void digit_reverse(int *, unsigned short *, unsigned short *, int);
void radix4(int, short[], short[]);
void fft();

RTDX_CreateInputChannel(cinput);
RTDX_CreateOutputChannel(coutput);

void main()
{
    int i,k;
    short tr[NUMPOINTS], ti[NUMPOINTS];

    // Target initialization for RTDXT
    TARGET_INITIALIZE();

    /*enable RTDX channels*/
    RTDX_enableInput(&cinput);
    RTDX_enableOutput(&coutput);

    //Read Twiddle factors to COMPLEX array and make Q-15;
    make_q15(tr, TR, NUMPOINTS); //Data in Header files from Matlab
    make_q15(ti, TI, NUMPOINTS);

    for(i=0; i<NUMPOINTS; i++)
    {
        W4[i].real = tr[i];
        W4[i].imag = ti[i];
    }

    /* Initialize A,B, IA, and IB arrays */
    for(k=0; k<NUMPOINTS; k++)
    {
    A[k].imag  = (short)(16383.0 * (-cos(2*PI/(double)(2*NUMPOINTS)*
            (double)k)));
```

```
      A[k].real =(short)(16383.0*(1.0 - sin(2*PI/(double)
               (2*NUMPOINTS)*(double)k)));
      B[k].imag =(short)(16383.0*(cos(2*PI/(double)(2*NUMPOINTS)*
               (double)k)));
      B[k].real =(short)(16383.0*(1.0 + sin(2*PI/(double)
               (2*NUMPOINTS)*(double)k)));
   }

   /* Initialize tables for FFT digit reversal function */
   R4DigitRevIndexTableGen(NUMPOINTS,&count,IIndex,JIndex);

   for(;;)    /* Infinite message loop. */
   {
      while(!RTDX_read(&cinput, g, sizeof(g)));

      /* Call FFT algorithm */
      fft();

      for (k=0; k<NUMDATA; k++)
      {
         magR[k] = (G[k].real*G[k].real) << 1;
         magI[k] = (G[k].imag*G[k].imag) << 1;

         output[k] = magR[k] + magI[k];
      }

      /* Write scaled data back to host. */
      RTDX_write(&coutput, &output, sizeof(output));
   }
}

void fft()
{
   int n;
   /* Forward DFT */
   /* From the 2N point real sequence, g(n), for the N-point
complex sequence, x(n) */

   for (n=0; n<NUMPOINTS; n++)
   {
      x[n].imag = g[2*n + 1]; /* x2(n) = g(2n + 1) */
      x[n].real = g[2*n]; /* x1(n) = g(2n) */
   }

   /* Compute the DFT of x(n) to get X(k) -> X(k) = DFT{x(n)} */
   radix4(NUMPOINTS, (short *)x, (short *)W4);
   digit_reverse((int *)x, IIndex, JIndex, count);
```

```
    /* Because of the periodicity property of the DFT, we know that
X(N+k) = X(k) . */
    x[NUMPOINTS].real = x[0].real;
    x[NUMPOINTS].imag = x[0].imag;
    /* The split function performs the additional computations
required to get G(k) from X(k). */

    split1(NUMPOINTS, x, A, B, G);
    /* Use complex conjugate symmetry properties to get the rest
of G(k) */
    G[NUMPOINTS].real = x[0].real - x[0].imag;
    G[NUMPOINTS].imag = 0;

    for (n=1; n<NUMPOINTS; n++)
    {
        G[2*NUMPOINTS-n].real = G[n].real;
        G[2*NUMPOINTS-n].imag = -G[n].imag;
    }
}

void radix4(int n, short x[], short w[])
{
    int n1, n2, ie, ia1, ia2, ia3, i0, i1, i2, i3, j, k;
    short t, r1, r2, s1, s2, co1, co2, co3, si1, si2, si3;
    n2 = n;
    ie = 1;
    for (k=n; k>1; k>>=2)
    {
        n1 = n2;
        n2 >>= 2;
        ia1 = 0;
        for (j=0; j<n2; j++)
        {
            ia2 = ia1 + ia1;
            ia3 = ia2 + ia1;
            co1 = w[ia1 * 2 + 1];
            si1 = w[ia1 * 2];
            co2 = w[ia2 * 2 + 1];
            si2 = w[ia2 * 2];
            co3 = w[ia3 * 2 + 1];
            si3 = w[ia3 * 2];
            ia1 = ia1 + ie;
            for (i0=j; i0<n; i0+=n1)
            {
                i1 = i0 + n2;
                i2 = i1 + n2;
                i3 = i2 + n2;
                r1 = x[2 * i0] + x[2 * i2];
                r2 = x[2 * i0] - x[2 * i2];
```

```
                    t = x[2 * i1] + x[2 * i3];
                    x[2 * i0] = r1 + t;
                    r1 = r1 - t;
                    s1 = x[2 * i0 + 1] + x[2 * i2 + 1];
                    s2 = x[2 * i0 + 1] - x[2 * i2 + 1];
                    t = x[2 * i1 + 1] + x[2 * i3 + 1];
                    x[2 * i0 + 1] = s1 + t;
                    s1 = s1 - t;
                    x[2 * i2] = (r1 * co2 + s1 * si2) >> 15;
                    x[2 * i2 + 1] = (s1 * co2-r1 * si2)>>15;
                    t = x[2 * i1 + 1] - x[2 * i3 + 1];
                    r1 = r2 + t;
                    r2 = r2 - t;
                    t = x[2 * i1] - x[2 * i3];
                    s1 = s2 - t;
                    s2 = s2 + t;
                    x[2 * i1] = (r1 * co1 + s1 * si1) >>15;
                    x[2 * i1 + 1] = (s1 * co1-r1 * si1)>>15;
                    x[2 * i3] = (r2 * co3 + s2 * si3) >>15;
                    x[2 * i3 + 1] = (s2 * co3-r2 * si3)>>15;
            }
        }
        ie <<= 2;
    }
}

void digit_reverse(int *yx, unsigned short *JIndex, unsigned short *
IIndex, int count)
{
    int i;
    unsigned short I, J;
    int YXI, YXJ;
    for (i=0; i<count; i++)
    {
        I = IIndex[i];
        J = JIndex[i];
        YXI = yx[I];
        YXJ = yx[J];
        yx[J] = YXI;
        yx[I] = YXJ;
    }
}

void R4DigitRevIndexTableGen(int n, int *count, unsigned short
*IIndex, unsigned short *JIndex)
{
    int j, n1, k, i;
    j = 1;
    n1 = n - 1;
```

```c
    *count = 0;
    for(i=1; i<=n1; i++)
    {
        if(i<j)
        {
            IIndex[*count] = (unsigned short)(i-1);
            JIndex[*count] = (unsigned short)(j-1);
            *count = *count + 1;
        }
        k = n >> 2;
        while(k*3 < j)
        {
            j = j - k*3;
            k = k >> 2;
        }
        j = j + k;
    }
}
void split1(int N, COMPLEX *X, COMPLEX *A, COMPLEX *B, COMPLEX *G)
{
    int k;
    int Tr, Ti;

    for (k=0; k<N; k++)
    {
        Tr = (int)X[k].real * (int)A[k].real - (int)X[k].imag *
             (int)A[k].imag+(int)X[N-k].real * (int)B[k].real +
             (int)X[N-k].imag * (int)B[k].imag;
        G[k].real = (short)(Tr>>15);

        Ti = (int)X[k].imag * (int)A[k].real + (int)X[k].real *
             (int)A[k].imag+(int)X[N-k].real * (int)B[k].imag -
             (int)X[N-k].imag * (int)B[k].real;
        G[k].imag = (short)(Ti>>15);
    }
}
void make_q15(short out[], float in[], int N)
{
    int i;

    for(i=0; i<N; i++)
    {
        out[i]=0x7fff*in[i];//Convert to Q-15, good approximate
    }
}
```

Care must be taken to avoid overflows if the algorithm is running on a fixed-point DSP. The amplitude of the input signal needs to be scaled properly in order to avoid overflows in the computations of the FFT on the DSP. In the example shown in Figure L9-14 256 is used as the amplitude of the input signal. Figure L9-15 illustrates an overflowed FFT outcome when the amplitude is set to 4096.

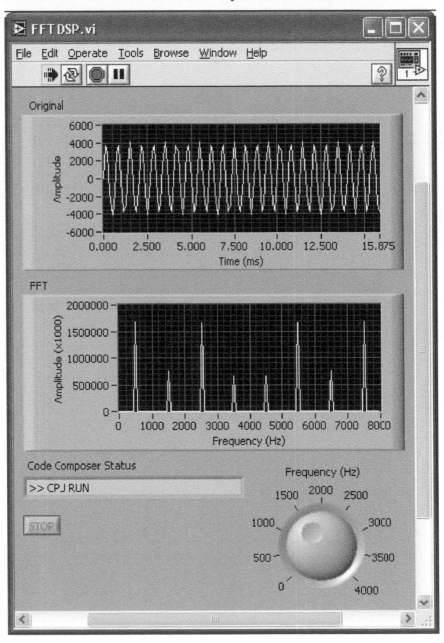

Figure L9-15: Overflow in computing FFT.

It should be noted that the overall timing of a typical DSP integration is often not so efficient due to the overhead associated with the RTDX communication. Nevertheless, the discussed DSP integration allows one to examine code execution on the C6x DSP hardware platform.

L9.6 Bibliography

[1] Texas Instruments, *Application Report SPRA 291*, 1997.

CHAPTER 10

DSP System Design: Dual Tone Multi-Frequency (DTMF) Signaling

In this and the next two chapters, three DSP system project examples are discussed and built using the LabVIEW hybrid programming approach. These examples show how relatively complex DSP systems can be devised in a relatively short amount of time by deploying hybrid programming. In the order of complexity, these examples consist of dual tone multi-frequency signaling, software-defined radio, and cochlear implant simulator.

Dual tone multi-frequency (DTMF) signaling is extensively used in voice communication applications such as voice mail and telephone banking. A DTMF signal is made up of two tones selected from a low and a high tone group. Each pair of tones contains one frequency from the low group (697 Hz, 770 Hz, 852 Hz, 941 Hz) and one frequency from the high group (1209 Hz, 1336 Hz, 1477 Hz). Figure 10-1 shows the frequencies allocated to the telephone pad push buttons.

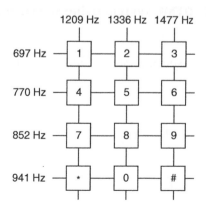

Figure 10-1: Keypad and allocated frequencies.

The implementation of the DTMF receiver system is normally done by using the Goertzel algorithm [1]. This algorithm is more efficient than the FFT algorithm for DTMF detection both in terms of the number of operations and amount of memory usage. Furthermore, unlike FFT, it does not require access to the entire data frame, leading to faster execution. As indicated in Figure 10-2, seven Goertzel filters are used here in parallel to form a DTMF detection system. Each Goertzel filter is designed to detect a DTMF tone. The output from each filter is squared and fed into a threshold detector, where the strongest signals from the low and high frequency groups are selected to identify a pressed digit on the keypad.

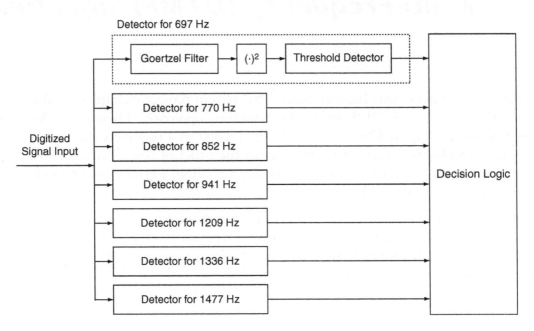

Figure 10-2: DTMF system using Goertzel algorithm.

The difference equations of a second-order Goertzel filter, as illustrated in Figure 10-3, are given by

$$v_k[n] = 2cos\left[\frac{2\pi k}{N}\right]v_k[n-1] - v_k[n-2] + x[n], \quad n = 0, 1, \ldots, N \qquad (10.1)$$

$$y_k[n] = v_k[n] - W_N^k v_k[n-1] \qquad (10.2)$$

where $x[n]$ denotes input, $y_k[n]$ denotes output, $v_k[n]$ denotes intermediate output, the subscript k indicates frequency bin, N is the DFT size, and $W_N^k = exp\left(-j\dfrac{2\pi k}{N}\right)$. The initial conditions are assumed to be zero, i.e., $v_k[-1] = v_k[-2] = 0$. Considering that only the magnitude of the signal is required for the DTMF tone detection, the following equation is used to generate magnitude squared outputs:

$$|y_k[N]|^2 = v_k^2[N] + v_k^2[N-1] - 2cos\left(\frac{2\pi k}{N}\right)v_k[N]v_k[N-1] \qquad (10.3)$$

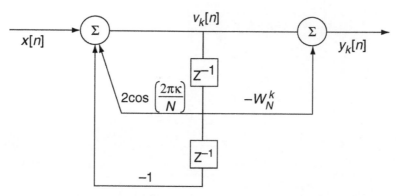

Figure 10-3: Structure of a second-order Goertzel filter.

The coefficients $2cos\left(\dfrac{2\pi k}{N}\right)$ are selected based on the DTMF tones. They are listed in Table 10-1.

Table 10-1: Fundamental Frequencies

Fundamental Frequency (Hz)	Coefficient
697	1.703275
770	1.635585
852	1.562297
941	1.482867
1209	1.163138
1336	1.008835
1477	0.790074

10.1 Bibliography

[1] E. Ifeachor and B. Jervis, *Digital Signal Processing: A Practical Approach*, Prentice-Hall, 2001.

Lab 10: Hybrid Programming of Dual Tone Multi-Frequency System

In this lab, a DTMF system is built by using the LabVIEW hybrid programming approach.

L10.1 DTMF Tone Generator System

Before building a BD for the DTMF system, let us begin by creating a keypad, as shown in Figure L10-1. The 12 buttons shown are grouped into a cluster so that their outputs are wired via a cluster wire. This is done by placing a Cluster shell (**Controls » Modern » Array, Matrix & Cluster » Cluster**) on the FP and then by locating Text Buttons (**Controls » Express » Buttons & Switches » OK Button)** in the cluster area.

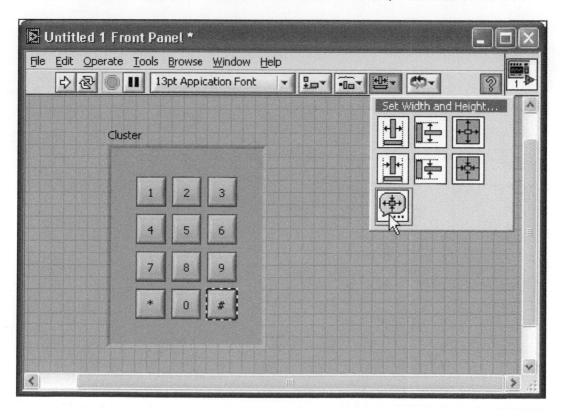

Figure L10-1: Creating a cluster control.

269

The width and height of a button can be adjusted for having a larger display on the FP. This is achieved by choosing the buttons and then selecting the option **Set Width and Height** from the **Resize Objects** menu of the FP toolbar, as shown in Figure L10-1. In this example, both the width and height of the buttons are set to 30. Also, the mechanical action of the buttons is configured to be **Latch When Released** by right-clicking on the buttons and choosing **Mechanical Action** from the shortcut menu. Once the configuration of a button is complete, the button is copied multiple times to construct a keypad. Change the Boolean text of the buttons appropriately. Align and distribute the buttons via **Align Objects** and **Distribute Objects** on the FP toolbar.

Next, the output value of the cluster control is specified for each button, i.e., the value coming out of the output of the cluster control when one of the buttons is pressed. To accomplish this, right-click on the border of the cluster and choose **Reorder Controls In Cluster** from the shortcut menu. This brings up the window shown in Figure L10-2. The numbers shown in the black background correspond to the modified order and the numbers in the white background to the original order.

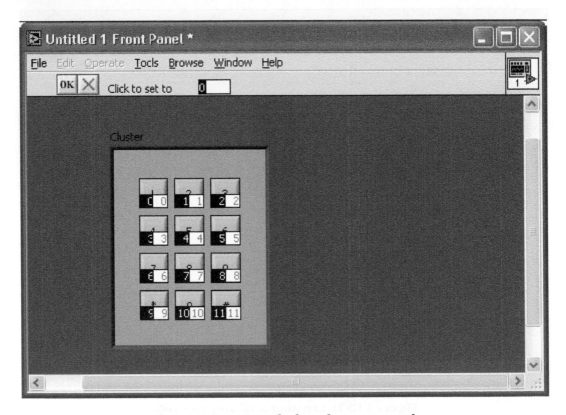

Figure L10-2: Reordering cluster control.

The number assigned to a key is displayed next to **Click to set to** shown on the toolbar. Click the buttons in a sequential order to specify the value shown in the toolbar area. After finishing the assignment of the values to the buttons, click the **OK** button to finish reordering controls or the **X** button to cancel the changes.

Right-click on the border of the cluster and choose **Auto Sizing » Size to Fit** to resize the cluster, if desired. Also, rename the label of the cluster as Keypad.

The BD of the built DTMF system is shown in Figure L10-3. Note that the keypad cluster control is shown as an icon on the BD. This VI generates a tone depending on the number pressed on the keypad in the FP. The DTMF decoder based on the Goertzel algorithm can be seen in the lower half of the BD.

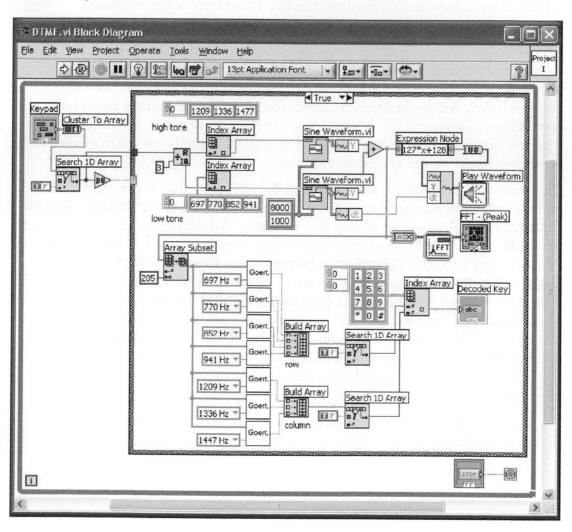

Figure L10-3: BD of the DTMF system.

To build the BD, wire the output value of the cluster to an array by using the `Cluster to Array` function (**Functions » Programming » Cluster & Variant » Cluster to Array**). This is done in order to have the value of each button as an element of an array. The array is then wired to a `Search 1D Array` function to search for the `True` values among the array elements. In other words, this is done to check the status of the buttons considering that the index of the array, which is greater than or equal to zero, is returned when a button is pressed; otherwise, −1 is returned.

Thus, if the index of the array becomes greater than or equal to zero, i.e., any button is pressed, a DTMF signal is generated and the decoding part in the `True` case of the `Case Structure` is executed. In the `False` case of the `Case Structure`, a time delay is included to continue the idle status until a key is pressed.

Now, let us go through the DTMF signal generation for the `True` case of the `Case Structure`. The value of the array index is wired to the `Quotient & Remainder` function with 3 as divisor. Since the numbers on the keypad are arranged in 3 columns and 4 rows, the remainder of this operation becomes the column index, and the quotient becomes the row index. Based on the column and row indices, a high and a low tone value are chosen using two 1D array constants. The low and high tone values are wired to a `Sine Waveform` VI to generate a waveform based on the chosen frequencies.

The generated waveform is scaled to an 8-bit integer so that it can be played at an audible volume level. An `Expression Node` (**Functions » Programming » Numeric » Expression Node**) is used for scaling the waveform. An `Expression Node` is useful for evaluating a simple equation or expression containing a single variable [1]. A `Play Waveform` VI (**Functions » Programming » Graphics & Sound » Sound » Output » Play Waveform**) is located to send out the waveform to the PC sound card. This is an Express VI which plays data from the sound output device based on a finite number of samples. The `Get Waveform Components` and `Build Waveform` VIs are used to build the waveform of the generated samples. For the spectral analysis of the generated samples, a `Spectral Measurement` Express VI and a waveform graph are used. The `Spectral Measurement` Express VI is configured as linear amplitude spectrum with no windowing.

At this stage, the DTMF generator is complete. The next section covers the decoding module.

L10.2 DTMF Decoder System

The Goertzel algorithm is used for the decoding of DTMF signals. The BD shown in Figure L10-4 illustrates the Goertzel algorithm described by Equation (10.3) using a MATLAB Script Node.

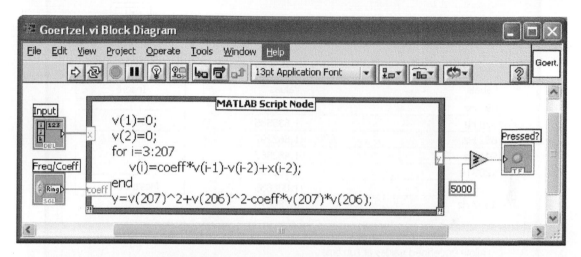

Figure L10-4: BD of Goertzel algorithm.

The inputs of this subVI consist of a 1D array of 205 samples and the coefficients of the Goertzel algorithm. A Text Ring control (**Controls » Modern » Ring & Enum » Text Ring**), labeled as Freq/Coeff, is located on the FP. Its data representation and properties are then modified as illustrated in Figure L10-5. This is done by right-clicking on the Text Ring in the FP and then choosing **Edit Items**... from the shortcut menu. Note that this Goertzel subVI is designed to incorporate only the first harmonic. It uses 205 input samples, and its coefficients are calculated based on 205 frequency bins.

The output of the Goertzel subVI is a Boolean value which indicates whether the specified frequency component is present in the input samples. This is decided by comparing the squared output of this subVI with a threshold value. The threshold value here is empirically set to 5000.

In the BD of the DTMF system shown in Figure L10-3, a total of seven Goertzel subVIs are placed to detect each frequency of a DTMF signal. The outputs of the Goertzel subVIs are grouped into two arrays to incorporate the row and column

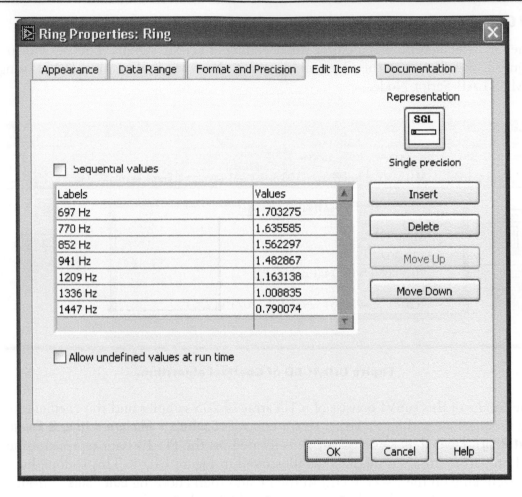

Figure L10-5: Ring properties.

frequencies. From these arrays, indices of the True values are searched to determine a pressed key. A string constant gets referred to by the indices of the 2D array of string constants.

To create the 2D array of string constants, first place an Array Constant shell (**Functions » Programming » Array » Array Constant**) on the BD. Then, place a String Constant (**Functions » Programming » String » String Constant**) in the Array Constant. As a result, a 1D array of string constants is created. In order to increase the dimension of the array, right-click on the Array Constant and choose **Add Dimension** from the shortcut menu. Now, enter the corresponding strings in the 2D array.

The output of the DTMF is shown in Figure L10-6. Notice that, when the button # is pressed, two frequencies are observed at 941 Hz and 1477 Hz in the decoded output. Furthermore, the decoded output matches the expected outcome.

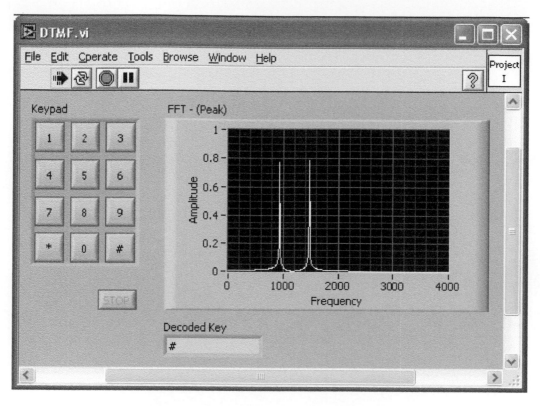

Figure L10-6: FP of the DTMF system.

L10.3 Bibliography

[1] National Instruments, *LabVIEW User Manual*, Part Number 320999E-01, 2003.

CHAPTER 11

DSP System Design: Software-Defined Radio

This chapter covers a software-defined radio system built by using the LabVIEW hybrid programming approach. A software-defined radio consists of a programmable communication system where functional changes can be made by merely updating software. For a detailed description of software-defined radio, the reader is referred to [1], [2].

4-QAM (Quadrature Amplitude Modulation) is chosen to be the modulation scheme of our software-defined radio system, noting that this modulation is widely used for data transmission applications over bandpass channels such as FAX modem, high-speed cable, multi-tone wireless, and satellite systems [2]. For simplicity, here the communication channel is considered to be ideal or noise-free.

11.1 QAM Transmitter

For transmission, pseudo noise (PN) sequences are generated to serve as our message signal. A PN sequence is generated with a five-stage linear feedback shift register structure, as shown in Figure 11-1, whose connection polynomial is given by

$$h(D) = 1 + D^2 + D^5 \qquad (11.1)$$

where D denotes delay and the summations represent modulo 2 additions.

The sequence generated by the preceding equation has a period of $31(= 2^5 - 1)$. Two PN sequence generators are used in order to create the message sequences for both the in-phase and quadrature phase components. The constellation of 4-QAM is shown in Figure 11-2. For more details of PN sequence generation, refer to [2].

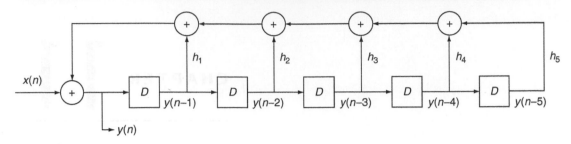

Figure 11-1: PN generation with linear feedback shift register.

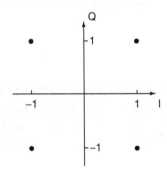

Figure 11-2: Constellation of 4-QAM.

Note that frame marker bits are inserted in front of the generated PN sequences. This is done for frame synchronization purposes, as discussed in the following receiver section. As illustrated in Figure 11-3, a total of 10 frame maker bits is located in front of each period of a PN sequence.

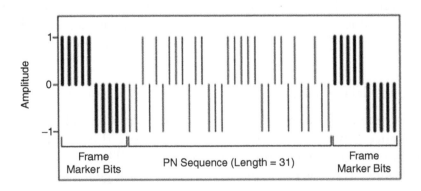

Figure 11-3: PN sequence generator.

The generated message sequences are then passed through a raised-cosine FIR filter to create a band-limited baseband signal. The frequency response of the raised cosine filter is given by

$$G(f) = \begin{cases} 1 & \text{for } |f| \leq (1 - \alpha)f_c \\ \cos^2\left[\dfrac{\pi}{4\alpha f_c}\left(|f| - (1 - \alpha)f_c\right)\right] & \text{for } (1 - \alpha)f_c \leq |f| \leq (1 + \alpha)f_c \\ 0 & \text{elsewhere} \end{cases} \quad (11.2)$$

where $\alpha \in [0, 1]$ denotes a roll-off factor specifying the excess bandwidth beyond the Nyquist frequency f_c.

The output of the raised cosine filter is then used to build a complex envelope, $\tilde{s}(t)$, of a QAM signal expressed by

$$\tilde{s}(t) = \sum_{k=-\infty}^{\infty} c_k\, g_T(t - kT) \quad (11.3)$$

where c_k indicates a complex message, made up of two real messages a_k and b_k, $c_k = a_k + jb_k$.

When $\tilde{s}(t)$ is modulated with $e^{j\omega_c t}$, an analytical signal or pre-envelope, $s_+(t)$, is generated,

$$s_+(t) = \tilde{s}(t)e^{j\omega_c t} = \sum_{k=-\infty}^{\infty} c_k\, g_T(t - kT)e^{j\omega_c t} \quad (11.4)$$

The transmitted QAM signal, $s(t)$, is thus given by

$$\begin{aligned} s(t) &= \Re[s_+(t)] \\ &= a(t)\cos(\omega_c t) - b(t)\sin(\omega_c t) \end{aligned} \quad (11.5)$$

where $\Re[\cdot]$ corresponds to the real part of the complex value inside the brackets.

Figure 11-4 illustrates the block diagram of the QAM transmitter just discussed. Notice that the two data paths, indicated by a solid line and a dotted line, represent complex data. Again, the reader is referred to [2] for more theoretical details.

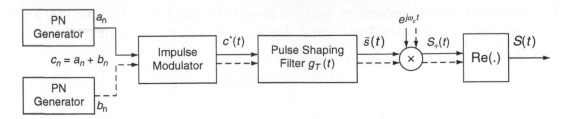

Figure 11-4: QAM transmitter [2].

11.2 QAM Receiver

11.2.1 Ideal QAM Demodulation

Here, it is assumed that the exact phase and frequency information of the carrier is available. The received QAM signal is denoted by $r(t)$. To simplify the system, an ideal channel is assumed between the transmitter and the receiver, i.e., $r(t) = s(t)$.

If $r(nT)$ is considered to be the sampled received signal, the analytic signal $r_+(nT)$ is given by

$$r_+(nT) = r(nT) + j\hat{r}(nT) \tag{11.6}$$

where $\hat{r}(\cdot)$ indicates the Hilbert transform of $r(\cdot)$. Thus, the complex envelope of the received QAM signal $\tilde{r}(nT)$ can be expressed as

$$\begin{aligned} \tilde{r}(nT) &= r_+(nT)e^{-j\omega_c nT} \\ &= a(nT) + jb(nT) \end{aligned} \tag{11.7}$$

Such a QAM demodulation process is illustrated in Figure 11-5.

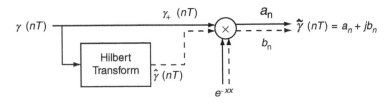

Figure 11-5: Ideal demodulation [2].

11.2.2 Frame Synchronization

Frame synchronization is required for properly grouping transmitted bits into an alphabet. To achieve this synchronization, a similarity measure, consisting of cross-correlation, is computed between the known marker bits and received samples. The cross-correlation of two complex values v and w is given by

$$R_{wv}[j] = \sum_{n=-\infty}^{\infty} \bar{w}[n]v[n+j] \qquad (11.8)$$

where the bar denotes complex conjugate.

An example of the cross-correlation outcome for frame synchronization is shown in Figure 11-6. The maximum value is found to be at the location of index 33. The subsequent message symbols are then framed from this index point.

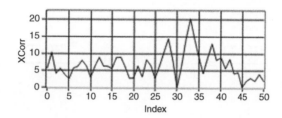

Figure 11-6: Cross-correlation of frame marker bits and received samples.

11.2.3 Decision-Based Carrier Tracking

Let us now consider the phase offset, denoted by θ, between the transmitter and the receiver. Based on this offset, the received signal can be written as

$$\tilde{r}(nT) = r_+(nT)e^{-j(\omega_c nT+\theta)}$$
$$= \hat{c}_n e^{-j\theta} \qquad (11.9)$$

where $\hat{c}_n$ indicates the output of a slicer mapping a received sample to the nearest ideal reference in the signal constellation. As a result, the baseband error at the receiver is given by

$$\tilde{e}(nT) = \hat{c}_n - \tilde{r}(nT) \qquad (11.10)$$

Next, the LMS update method is used to minimize a decision-directed cost function, $J_{DD}(\theta)$, consisting of the mean squared baseband error

$$J_{DD}(\theta) = avg\left[|\tilde{e}(nT)|^2\right]$$

$$= avg\left[\tilde{e}(nT)\overline{\tilde{e}(nT)}\right] \tag{11.11}$$

By differentiating $J_{DD}(\theta)$ with respect to θ, we get

$$\frac{dJ_{DD}(\theta)}{d\theta} = avg\left[\frac{d[\tilde{e}(nT)\overline{\tilde{e}(nT)}]}{d\theta}\right]$$

$$= 2avg\left[\Re\left\{\overline{\tilde{e}(nT)}\frac{d\tilde{e}(nT)}{d\theta}\right\}\right] \tag{11.12}$$

where

$$\frac{d\tilde{e}(nT)}{d\theta} = \frac{d}{d\theta}[\hat{c}_n - \tilde{r}(nT)] = -\frac{d\tilde{r}(nT)}{d\theta} \tag{11.13}$$

and

$$\frac{d\tilde{r}(nT)}{d\theta} = -j\hat{c}_n e^{-j\theta} = -j\tilde{r}(nT) \tag{11.14}$$

Equation (11.12) can thus be rewritten as

$$\frac{dJ_{DD}(\theta)}{d\theta} = 2avg\left[\Re\left\{\overline{\tilde{e}(nT)}j\tilde{r}(nT)\right\}\right]$$

$$= -2avg\left[\Im\left\{\overline{\tilde{e}(nT)}\tilde{r}(nT)\right\}\right] \tag{11.15}$$

$$= -2avg\left[\Im\left\{\overline{\hat{c}_n}\tilde{r}(nT)\right\}\right]$$

where $\Im[\cdot]$ corresponds to the imaginary part of the complex value inside the brackets.

By writing the term $\Im\{\overline{\tilde{c}_n}\tilde{r}(nT)\}$ in polar form, we get

$$\Im\{\overline{\tilde{c}_n}\tilde{r}(nT)\} = \Im\{\overline{R_c e^{j\beta_c}}R_r e^{j\beta_r}\}$$
$$= R_c R_r \sin(\beta_r - \beta_c) \tag{11.16}$$

Thus,

$$\sin(\beta_r - \beta_c) = \frac{\Im\{\overline{\tilde{c}_n}\tilde{r}(nT)\}}{R_c R_r} \tag{11.17}$$

Note that for small $\beta_r - \beta_c$,

$$\sin(\beta_r - \beta_c) \approx \beta_r - \beta_c$$
$$R_r \approx R_c = |c_n| \tag{11.18}$$

As a result, the phase error $\Delta\theta(n)$ is given by

$$\Delta\theta(n) = \frac{\Im\{\overline{\tilde{e}(nT)}\tilde{r}(nT)\}}{|c_n|^2} \tag{11.19}$$

Figure 11-7 shows a block diagram of the preceding tracking equations.

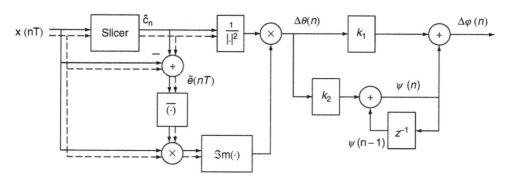

Figure 11-7: Decision-directed carrier phase and frequency tracking.

When both phase and frequency tracking are considered, the carrier phase of the receiver becomes

$$\phi(n+1) = \phi(n) + \Delta\phi(n) \tag{11.20}$$

In this case, the phase update $\Delta\phi(n)$ is given by

$$\Delta\phi(n) = k_1 \Delta\theta(n) + \psi(n) \tag{11.21}$$

where $\psi(n)$ denotes the contribution of frequency tracking, which is expressed as

$$\psi(n) = \psi(n-1) + k_2 \Delta\theta(n) \tag{11.22}$$

The scale factors k_1 and k_2 are configured to be small here, and usually $k_1/k_2 \geq 100$ is required for phase convergence [2].

11.3 Bibliography

[1] C. Johnson and W. Sethares, *Telecommunication Breakdown: Concepts of Communication Transmitted via Software-Defined Radio*, Prentice-Hall, 2004.

[2] S. Tretter, *Communication System Design Using DSP Algorithms*, Klumer Academic/Plenum Publishers, 2003.

Lab 11: Hybrid Programming of a 4-QAM Modem System

The design of a 4-QAM modem system is covered in this lab. As shown in Figure L11-1, this system consists of the following functional modules: message source, pulse shape filter, QAM modulator, Hilbert transformer, QAM demodulator, frame synchronizer, and phase and frequency tracker. The system is divided into two parts: transmitter and receiver. The first three modules (message source, pulse shape filter, and QAM modulator) make up the transmitter side; and the other modules, the receiver side. The building of each functional module is described in the sections that follow.

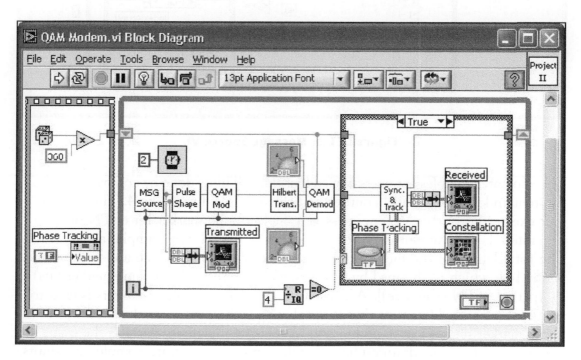

Figure L11-1: System-level VI of 4-QAM modem.

L11.1 QAM Transmitter

The first component of the QAM modem is the message source. Here, PN sequences are used for this purpose. Frame marker bits are inserted in front of these sequences to achieve frame synchronization. The BD of the Message Source VI is shown in Figure L11-2.

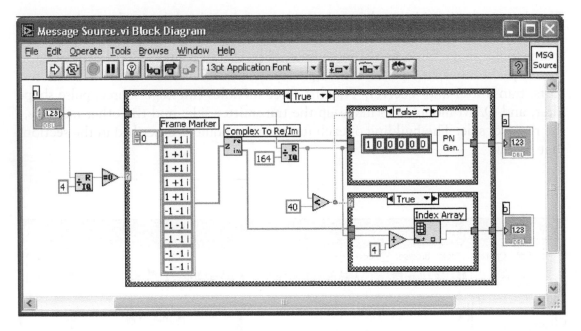

Figure L11-2: Message Source VI.

The generated samples are oversampled four times. This is done by comparing with 0 the remainder of the global counter, indicated by n, divided by 4. Thus, out of four executions of this VI, one message sample (frame marker bit or PN sample) is generated. For the remaining three executions of the VI, zero samples are generated. The total length of the message for one period of a PN sequence and frame marker bits is 164, which is obtained by 4 (oversampling rate) × [10 (frame marker bits) + 31 (period of PN sequence)]. A constant array of 10 complex numbers is used to specify the marker bits. Note that the real parts of the complex values are used as the frame marker bits of the in-phase samples and the imaginary parts as the frame marker bits of the quadrature-phase samples. In order to create complex constants, one needs to change the representation of a numeric constant by right-clicking on it and choosing **Representation » Complex Double** (or **Complex Single**).

The BD of the PN Generator VI is shown in Figure L11-3. With this subVI, a pseudo noise sequence of length 31 is generated by XORing the values of the second and fifth shift registers.

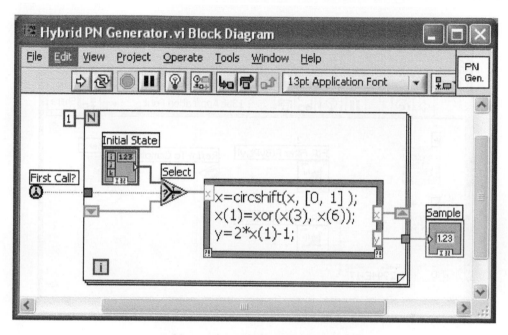

Figure L11-3: PN Generator VI.

The Shift Register, Select, and MathScript Node VIs are used to compute a new PN sample. A For Loop with one iteration and a First Call? function (**Functions » Programming » Synchronization » First Call?**) are used to pass the shift register value of a current call to a next call of the subVI. The First Call? function checks whether a current call is occurring for the first time. If that is the case, the shift register values are initialized by their specified initial values. Otherwise, the old values of the shift registers are passed from the previous execution of the subVI. Notice that the PN Generator VI shown in Figure L11-3 is built with the consideration of porting the algorithm to a DSP hardware platform. Alternatively, the built-in Binary MLS VI (**Functions » Signal Processing » Signal Generation » Binary MLS**) can be used for the LabVIEW implementation.

Next, the generated samples are passed to a pulse shape filter shown in Figure L11-4. A raised cosine filter is used to serve as the pulse shape filter. The FIR Filter PtByPt VI is utilized for this purpose. The two outputs of the pulse shape filters are combined to construct the pulse-shaped message signal by using the Re/Im to Complex function (**Functions » Programming » Numeric » Complex » Re/Im to Complex**).

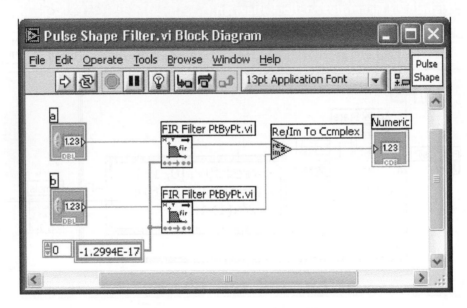

Figure L11-4: Pulse shape filter.

As for the filter coefficients, they can be designed by a filter design tool, such as the one discussed in Lab 4, and stored in an array of constants.

The signal passed through the pulse shape filter is then connected to the QAM modulator shown in Figure L11-5. The QAM modulated signal $s(t)$ is obtained by taking the real part of the pre-envelope signal $s_+(t)$. This is achieved by performing a complex multiplication between the complex input and a complex carrier consisting of a cosine and a sine waveform. This completes the modules of the transmitter. In the next section, the modules of the receiver are built.

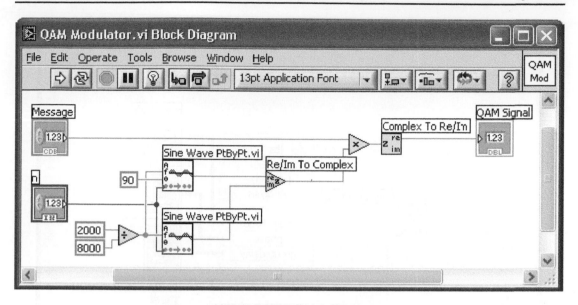

Figure L11-5: QAM modulator.

L11.2 QAM Receiver

The first module on the receiver side is the Hilbert transformer. This module builds the required analytic signal for demodulation based on the transmitted QAM signal.

A Hilbert transformer is built by using the DFD Remez Design VI (**Functions » Addons » Digital Filter Design » Filter Design » Advance FIR Filter Design » DFD Remez Design**) of the DFD toolkit. To have an integer group delay, an even number, such as 32, is specified as the filter order. The DFD Filter Analysis Express VI is wired to analyze the group delay of the filter as well as its magnitude and phase response, as shown in Figure L11-6.

The specifications of the Hilbert transformer are similar to those of a bandpass filter, as indicated in Figure L11-7. Notice that only one element of the cluster array is needed to design the Hilbert transformer. However, when a control is created at the band specs terminal of the DFD Remez Design VI, there are two default cluster values. The second element, indexed at 1, should thus be deleted. To do this, select the element of the cluster array to be deleted; then right-click and choose **Data Operation » Delete Element** from the shortcut menu.

By running the VI, one can see the magnitude, phase response, and group delay of the Hilbert transformer, as shown in Figure L11-7.

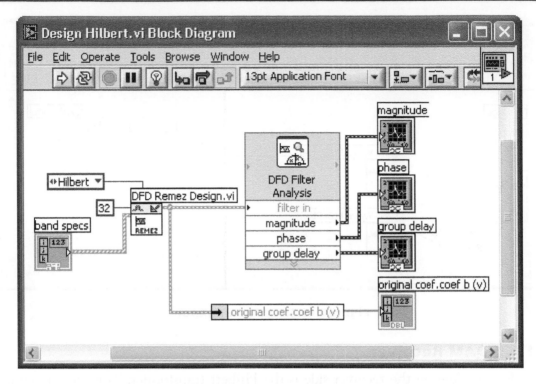

Figure L11-6: Building the Hilbert transformer.

The array of indicators corresponding to the Hilbert transform coefficients is converted to an array of constants to be used by the other VIs. Note that the design and analysis of the Hilbert transformer are needed only in the designing phase, not in the implementation phase of the modem system.

The BD of the Hilbert transformer using the coefficients obtained from the DFD toolkit is shown in Figure L11-8.

A `Data Queue PtByPt` VI (**Functions » Signal Processing » Point By Point » Other Functions PtByPt » Data Queue PtByPt**) is employed in order to synchronize the input and output of the Hilbert transformer. In other words, the input samples are delayed until the corresponding output samples become available. This is needed due to the group delay associated with the filtering operation. For an FIR filter of 33 taps, the group delay is 16. An array of numeric constants corresponding to the filter coefficients is set up based on the text file generated by a filter design tool. Here, an FIR filter has been used for the implementation of the Hilbert transformer instead of the built-in VI of LabVIEW. This is done to allow its execution by a DSP platform.

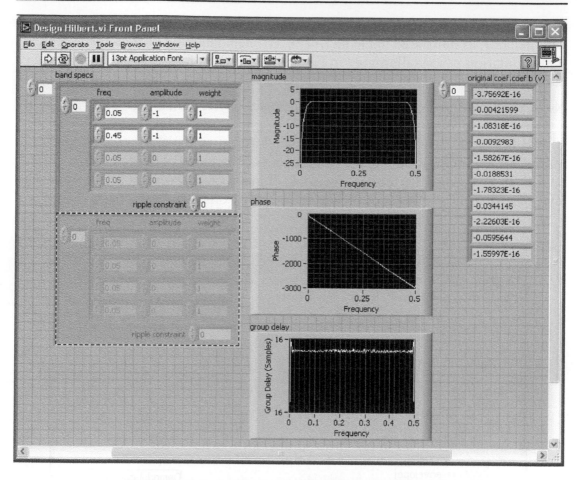

Figure L11-7: Analysis of the Hilbert transformer.

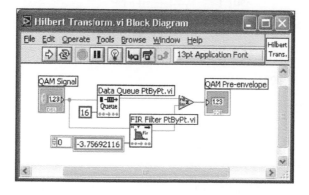

Figure L11-8: Hilbert Transform VI.

The analytic signal achieved from the Hilbert transformer is demodulated by the QAM demodulator, as illustrated in Figure L11-9. The demodulation process is similar to the modulation process illustrated in Figure L11-5 except for the negative frequency part.

Next, the QAM demodulated signal is decimated by 4. To do this, one can use a Case Structure so that every fourth sample is selected for processing, as illustrated in Figure L11-1. The decimated signal is sent to the Sync & Tracking VI for frame synchronization and phase/frequency tracking. The Sync & Tracking VI is an intermediate-level subVI incorporating several subVIs/functions and operating in two different modes: frame synchronization and phase/frequency

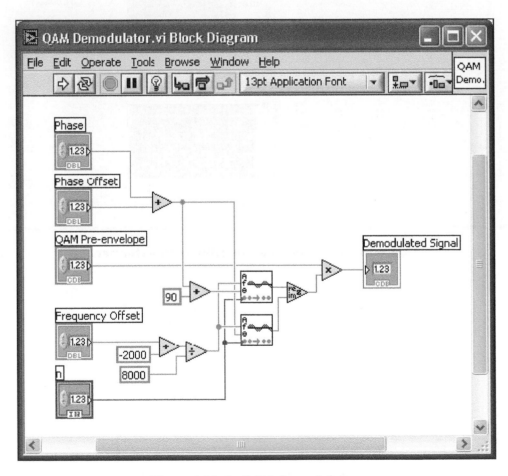

Figure L11-9: QAM demodulator.

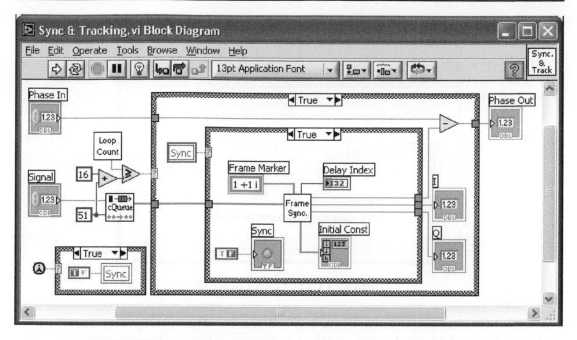

Figure L11-10: Sync & Tracking VI—frame synchronization mode.

tracking. Let us examine the BD of this VI displayed in Figure L11-10. The input samples are passed into the receiver queue, implemented via the `Complex Queue PtByPt` VI (**Functions » Signal Processing » Point By Point » Other Functions PtByPt » Complex Queue PtByPt**), in order to obtain the beginning of a frame by cross-correlating the frame marker bits and received samples in the queue. The queue continues to be filled until it is completely full. Extra iterations are done to avoid including any transient samples due to the delays associated with the filtering operations in the transmitter.

The length of the queue is configured to be 51 in order to include the entire marker bits in the queue. This length is decided based on this calculation: 31 (one period of PN sequence) + 2 × 10 (frame marker bits). Also, 16 extra samples are taken to flush out any possible transient output of the filter, as mentioned previously. Bear in mind that the length of the queue or the number of extra reads varies based on the specification of the transmitted signal, such as the length of the frame marker bits and the number of taps of the phase shape filter. A counter, denoted by the `Loop Count` VI in Figure L11-11, is used to count the number of samples filling the queue. Once the queue is completely filled and extra reads are done, the frame synchronization module is initiated.

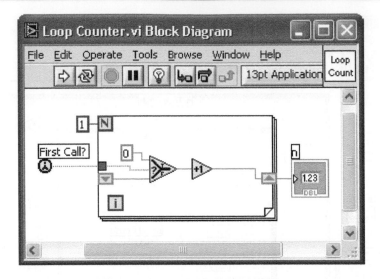

Figure L11-11: Loop Count VI.

The subVI for frame synchronization implemented within a MathScript Node is shown in Figure L11-12. In this subVI, the cross-correlation of the frame marker bits and the samples in the receiver queue are computed. Lines 1 to 4 perform the

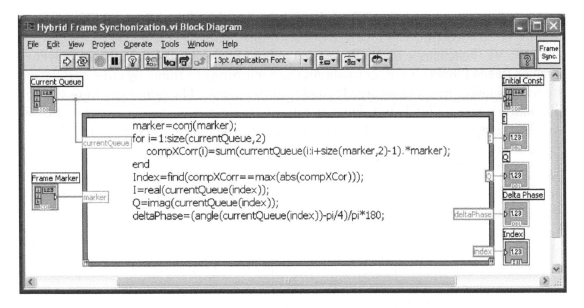

Figure L11-12: Frame Synchronization VI.

complex cross-correlation operation as indicated in Equation (11.8). The absolute value of the complex correlation output is used to obtain the cross-correlation peak, since the location of this peak coincides with the beginning of the frame. The find function in line 5 is used to detect the index corresponding to the maximum cross-correlation value.

Once the index of the maximum cross-correlation value is obtained, all the data samples are taken at this location of the queue. Consequently, the data bits get synchronized.

The initial phase estimation is achieved using the phase of the complex data at the beginning of the marker bits. Considering that the ideal reference is known for the first bit of the frame marker, $1+i$ in our case, this allows us to obtain the phase difference between the ideal reference and the received frame marker bits. The real and imaginary parts of data at the beginning of the marker bits are also passed to the Phase & Frequency Tracking VI to provide the initial constellation.

The subVI of the frame synchronization is now complete. Locate the subVI on the BD of the Sync & Tracking VI shown in Figure L11-10. Notice that three local variables are created in order to pass the indicator values to the other parts of the VI which cannot be wired. In the Sync & Tracking VI, a Rounded LED indicator, labeled as Sync, is placed on the FP. A local variable is created by right-clicking either on the terminal icon in the BD or on the Rounded LED indicator in the FP and choosing **Create » Local Variable**. Next, a local variable icon is placed on the BD. More details on using local and global variables can be found in [1].

The local variable Sync is used to control the flow of data for the frame synchronization. The initial value of the local variable is set to True to execute the frame synchronization. Then, it is changed to False within the Case Structure so that it is not invoked again. The other two local variables, Initial Const and Delay Index, are used as the inputs of the phase and frequency tracking module, as shown in Figure L11-13.

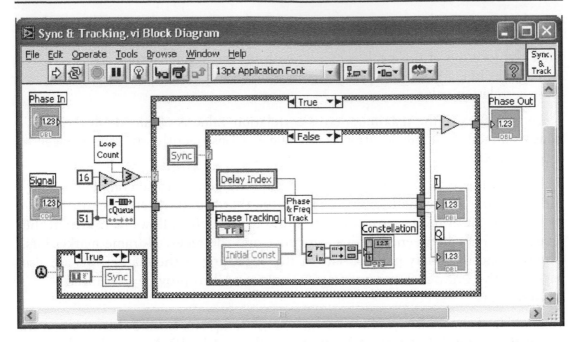

Figure L11-13: Sync & Tracking VI—phase and frequency tracking mode.

Now, let us describe the Phase & Frequency Tracking VI illustrated in Figure L11-14. A Formula Node (**Functions » Programming » Structures » Formula Node**) is shown in the upper part of the BD, which acts as a slicer to determine the nearest ideal reference based on the quadrant on the I-Q plane. A Formula Node structure is capable of evaluating a script written in text-based C code. There are numerous built-in mathematical functions and variables which can be used in a Formula Node. For example, pi represents π in the formula node script shown in Figure 11-21. Further details on Formula Node can be found in [1].

The phase error, as shown in the BD in Figure L11-14, is computed from Equation (11.19). This error is multiplied by a small-scale factor to determine the phase update $\Delta\varphi(n)$ in a second Formula Node implementing Equation (11.20).

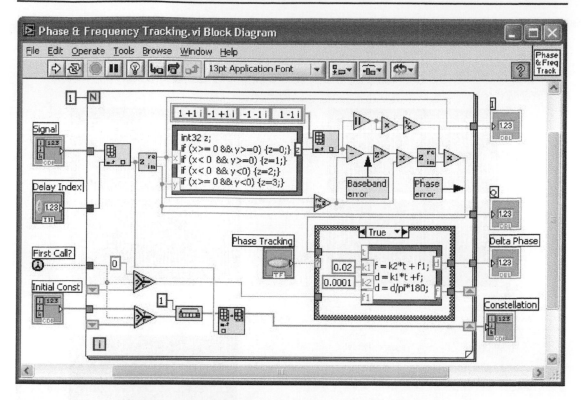

Figure L11-14: Phase & Frequency Tracking VI.

Now, all the components of the modem system are completed. As the final step, a waveform chart and an XY Graph (**Controls » Modern » Graph » XY Graph**) are added to the system-level BD shown in Figure L11-1. Figure L11-15 shows the FP of the system. After the phase is updated, the received signal becomes nearly a perfect reproduction of the transmitted signal except for the time delay. If there exist a phase and a frequency offset with no tracking, the received signal appears as shown in Figure L11-16. As displayed in this figure, the constellation of the received signal is rotated, and the amplitudes of some of the received samples become too small. Obviously, the received signal will change by introducing channel noise.

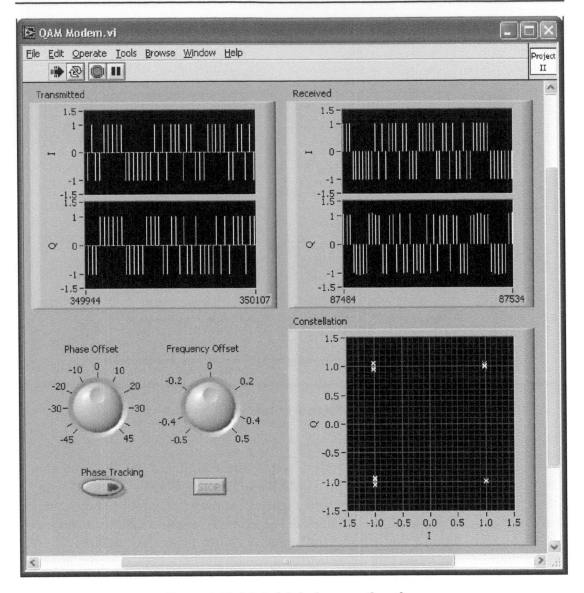

Figure L11-15: Initial phase estimation.

The change in the constellation via the phase and frequency tracking is illustrated in Figure L11-17. The constellation of the samples in the I-Q plane becomes that of the ideal reference as the tracking operation progresses.

In summary, a 4-QAM transmitter and receiver system is built in LabVIEW by adopting a hierarchical approach. A simplified version of the system hierarchy, displayed by choosing **View » VI Hierarchy**, is shown in

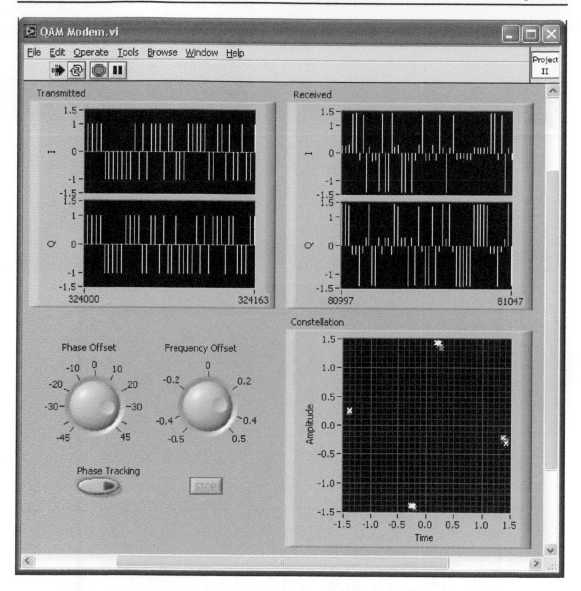

Figure L11-16: Received signal with no phase and frequency tracking.

Figure L11-18. When the phase and frequency tracking module is used, the phase and/or frequency offset between the transmitter and receiver is successfully compensated.

Note that all the subVIs discussed in this lab can be saved in a LabVIEW Library (LLB) file, e.g., *Lab 11.llb*. A new LLB file can be created by choosing **New VI Library** and naming it from the Name the VI window, which is brought up during the save operation.

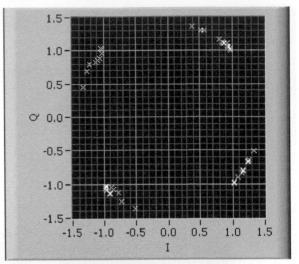

Figure L11-17: Phase and frequency tracking in the I-Q plane.

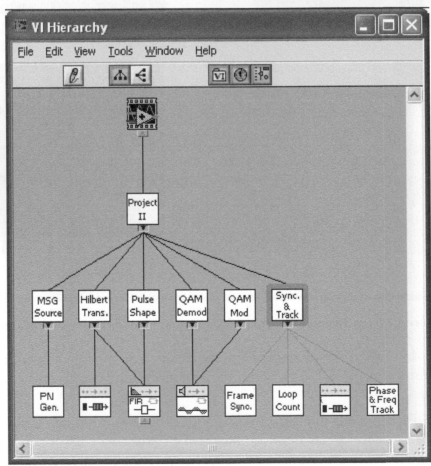

Figure L11-18: Hierarchy of the QAM Modem VI.

L11.3 Bibliography

[1] National Instruments, *LabVIEW User Manual*, Part Number 320999E-01, 2003.

11.3 Bibliography

[1] National Instruments, LabVIEW User Manual, Part Number 320999E-01, 2003.

12

DSP System Design: Cochlear Implant Simulator

This chapter covers the real-time implementation of a cochlear implant signal processing simulator by using the LabVIEW hybrid programming approach. A cochlear implant is a prosthetic device surgically implanted into the inner ear that is used to restore partial hearing in profoundly deaf people or patients suffering from nerve deafness. In this device, sounds acquired from a microphone are converted into electrical signals, which are then transmitted to a number of implanted electrodes in the cochlea via radio waves leading to hearing perception [1]. This chapter covers the signal processing components of a cochlear implant system and their implementations in a hybrid mode.

12.1 Cochlear Implant System

A cochlear implant system consists of the following four major components: (1) a microphone that picks up an input speech signal, (2) a signal processor that converts this signal into electrical signals, (3) a transmission system that transmits the electrical signals to implanted electrodes in the cochlea, and (4) an array of electrodes that are surgically inserted into the cochlea. Via the array of electrodes, auditory nerve fibers at different locations in the cochlea get stimulated depending on the signal frequency. A signal processor is used for bandpass filtering the input speech signal into several (12–22) frequency bands. The processor converts the signals from each band into electrical signals and delivers them to the array of electrodes. For a detailed description of cochlear implants, see [1].

Different signal processing strategies have been presented in the literature for converting acoustic signals to electrical signals [1], [2]. Here, a vocoder-based strategy known as Continuous Interleaved Sampling (CIS) [2] is considered. This vocoder strategy is widely used in commercial cochlear implants. In order to conduct a qualitative analysis of electrical stimuli obtained via the vocoder strategy, a synthesis stage is included along with the decomposition stage in our implementation, as illustrated in Figure 12-1. The synthesis method implemented here is based on the noise-band vocoder simulation reported in [2].

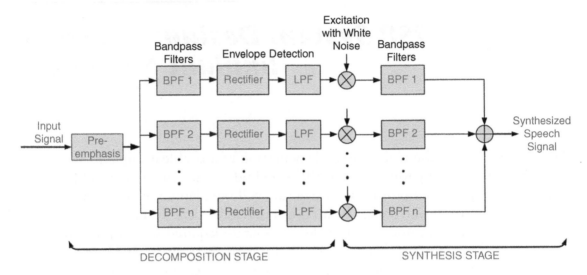

Figure 12-1: Noise-band vocoder simulation of cochlear implants.

As shown in Figure 12-1, during the decomposition stage, an input speech signal is first pre-emphasized and passed through a bank of bandpass filters. The cut-off frequencies for the bandpass filters are obtained by logarithmically dividing the speech signal bandwidth equally over a given number of channels. The envelopes of the filtered signals are then extracted via full-wave rectification and lowpass filtering with a typical cut-off frequency of 400 Hz. During the synthesis stage, the envelopes obtained after the decomposition are excited with white noise and then filtered through the same bank of bandpass filters that are used during the decomposition stage. Finally, a synthesized signal is reconstructed by summing all the filtered signals as indicated in Figure 12-1. In essence, the CIS strategy consists of the following signal processing parts: pre-emphasis filter, filterbank of bandpass filters for decomposition and synthesis stages, envelope detection (full-wave rectification + lowpass filtering), and white noise excitation.

12.2 Real-Time Implementation

In order to process acquired speech frames from a microphone in real-time, the coefficients of the filters need to be computed. The real-time implementation of the CIS strategy is done here in two phases: *Design* and *Real-Time* [3]. In the Design phase, the filter coefficients for the pre-emphasis filter, bandpass filters, and lowpass filter are computed based on user-specified characteristics including filter order and cut-off frequency. In addition, a white noise sequence is generated for noise excitation during the synthesis stage depending on the frame length specified by the user. In the Real-Time phase, all the computed filter coefficients together with the white noise sequence are fed back continuously within a real-time loop, as indicated in Figure 12-2.

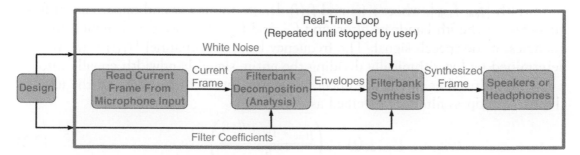

Figure 12-2: Real-time implementation flow of noise-band vocoder strategy.

In the real-time loop, an input frame is first acquired from a microphone at a given sampling rate. The pre-emphasized output is then passed through the filter bank decomposition and synthesis stages. Finally, the synthesized frames are sent to a speaker or headphone to provide audio feedback. This process is repeated continuously within the real-time loop until the user halts the process. Since this process is based on continuous acquisition and processing of input frames, having a real-time throughput is essential. For this purpose, it is required to complete the processing of a current frame before a next frame is captured. In other words, the total processing time for a frame should not exceed its length in time. More details of each of the CIS components follow.

12.2.1 Pre-Emphasis Filter

The pre-emphasis component consists of a first order IIR filter whose forward coefficients b_p and reverse coefficients a_p are given by

$$b_p = [1 - \exp(-1200 * 2 * pi/F_s)]$$

$$a_p = [1 - \exp(-3000 * 2 * pi/F_s)] \tag{12.1}$$

where F_s denotes the sampling frequency.

12.2.2 Filterbank for Decomposition and Synthesis

During the decomposition stage, the bandpass filter associated with each channel passes only a certain frequency band $[f_1(i)\ f_2(i)]$ of the entire input speech signal bandwidth $[f_{low}\ f_{high}]$, where $f_1(i)$ and $f_2(i)$ denote the lower and upper cut-off frequencies for the ith bandpass filter, and f_{low} and f_{high} the lowest and highest frequencies of the speech signal. The frequency band of a channel $[f_1(i)\ f_2(i)]$ is determined by logarithmically dividing the entire signal bandwidth equally into n channels. In other words, the upper and lower cut-off frequencies for the ith channel bandpass filter are specified as

$$f_1(i) = \left(\frac{f_{low} * 10^{C_b*(i-1)}}{F_s}\right) \tag{12.2}$$

$$f_2(i) = \left(\frac{f_{low} * 10^{C_b*(i)}}{F_s}\right) \tag{12.3}$$

where C_b denotes the channel bandwidth given by

$$C_b = \left(\frac{\log_{10}(f_{high}/f_{low})}{n}\right) \tag{12.4}$$

12.2.3 Envelope Detection

The full-wave rectified output of each channel is obtained by taking the absolute value of the corresponding bandpass filter output. This output is then passed through a lowpass filter having a cut-off frequency normally set at 400 Hz to extract its envelope.

12.2.4 White Noise Excitation

Finally, a white noise sequence is used to modulate the envelope of each channel. Here, a white noise sequence is generated by passing a uniformly distributed pseudorandom sequence with the values in the range [–1 1] through the *sgn* function defined as

$$\text{sgn}(x) = \begin{cases} -1, & x < 0 \\ 0, & x = 0 \\ 1, & x > 0 \end{cases} \tag{12.5}$$

12.3 Bibliography

[1] P. Loizou, "Mimicking the human ear," *IEEE Signal Processing Magazine*, vol. 15, pp. 101–130, 1998.

[2] P. Loizou, "Speech processing in vocoder-centric cochlear implants," *Cochlear and Brainstem Implants* (ed. Moller, A.), Adv. Otorhinolaryngol. Basel, Karger, vol. 64, pp. 109–143, 2006.

[3] V. Peddigari, N. Kehtarnavaz, and P. Loizou, "Real-time LabVIEW implementation of cochlear implant signal processing on PDA platforms," *Proceedings of IEEE Conference on Acoustics, Speech and Signal Processing*, vol. 2, pp. 357–360, 2007.

Lab 12: Hybrid Programming of Cochlear Implant Simulator System

This lab shows how a cochlear implant simulator can be built in a hybrid mode as a DSP system design application example. Figure L12-1 illustrates the system-level BD of such a system consisting of the following functional modules: filter design and real-time loop. The filter design module is executed only once before the real-time loop is initiated for the purpose of computing the pre-emphasis, bandpass, and lowpass filter coefficients depending on user-specified input parameters. Then, the processing is repeated continuously in a `While Loop` until the user halts the program by choosing the `Exit` control. It uses two `Case Structures`: one to initiate the real-time loop via a `Start` Boolean control and the other to stop the real-time loop via a `Stop` Boolean control. For each of these Boolean controls, first right-click and choose the **Properties** menu. Under the Properties menu, choose the **Operation** tab and set the **Button behavior** as **Latch when released**.

When the user presses the `Start` Boolean control, the `DesignFilterCoeff` VI computes the pre-emphasis, bandpass, and lowpass filter coefficients and passes them to a C DLL. The C DLL implements the system within a real-time loop using the CALLBACK function supported by Windows APIs [1]. Once the real-time loop is initiated via CALLBACK, the cochlear implant signal processing keeps running in the background till halted via the `Stop` Boolean control or the `Exit` control. When the `Stop` Boolean control is pressed, the real-time loop is halted, and the memory allocated to all global variables gets cleared.

As can be seen from Figure L12-1, a `Flat Sequence` Structure (**Functions » Programming » Structures » Flat Sequence Structure**) is used to initialize the `Processing Time (ms)` before initiating the cochlear implant simulator. The `Processing Time (ms)` indicates the amount of time which is required to obtain a synthesized frame for an input speech signal frame. More details on building the functional module `DesignFilterCoeff` subVI and the C DLL incorporating the functions `StartCISProcess` and `StopCISProcess` are presented next.

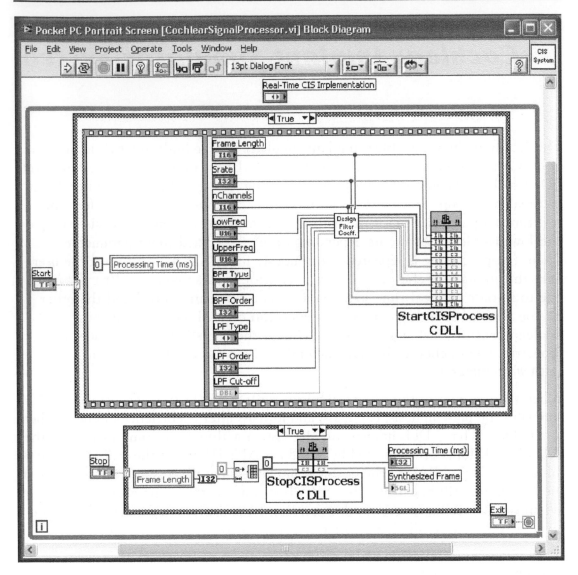

Figure L12-1: System-level VI of Cochlear Implant Simulator.

L12.1 Filter Design

The cochlear implant simulator presented here is built using the vocoder-based Continuous Interleaved Sampling (CIS) strategy. Figure L12-2 shows the BD of the DesignFilterCoeff VI, which computes the pre-emphasis filter, bandpass filter, and lowpass filter coefficients, along with the white noise sequence. As shown

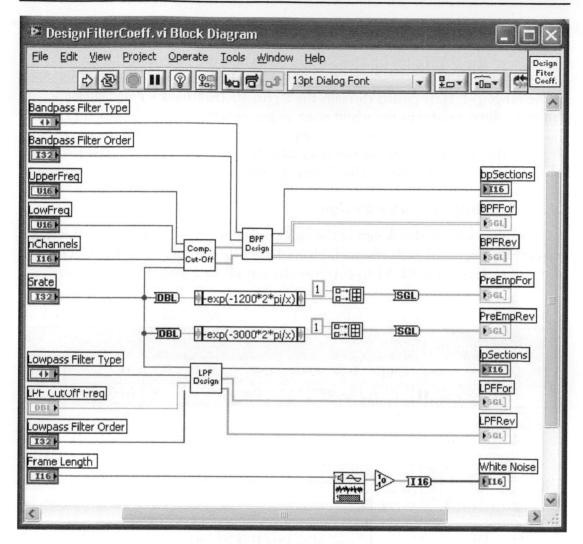

Figure L12-2: Filter Design VI.

in this figure, an Expression Node (**Functions » Mathematics » Numeric » Expression Node**) is used to compute the pre-emphasis filter coefficients given by Equation (12.1). The sampling frequency, Srate, is passed as an input to the Expression Node to evaluate the first-order pre-emphasis filter coefficients, which include both forward and reverse coefficients PreEmpFor and PreEmpRev, respectively.

Next, a white noise sequence, named White Noise, is generated using the Uniform White Noise VI (**Functions » Signal Processing » Signal Generation »**

Uniform White Noise). The length of the white noise sequence is equal to the number of samples in the acquired speech frame and is specified by passing the Frame Length to the Uniform White Noise VI. Then the output of the Uniform White Noise VI, which generates uniformly distributed pseudo random numbers in the range [–1 1], is passed through the Sign VI (**Functions » Programming » Numeric » Sign**) to obtain the white noise sequence according to Equation (12.5). It should be noted that this white noise sequence is used to modulate the envelopes obtained during the decomposition stage of CIS as discussed earlier. The following subsections elaborate more on the design of lowpass and bandpass filter coefficients.

L12.1.1 Bandpass Filter Design

As a first step toward the design of the bandpass filters, it is required to compute the cut-off frequencies of each channel. Figure L12-3 shows the system-level BD of the ComputeCutoff VI to compute the cut-off frequencies according to

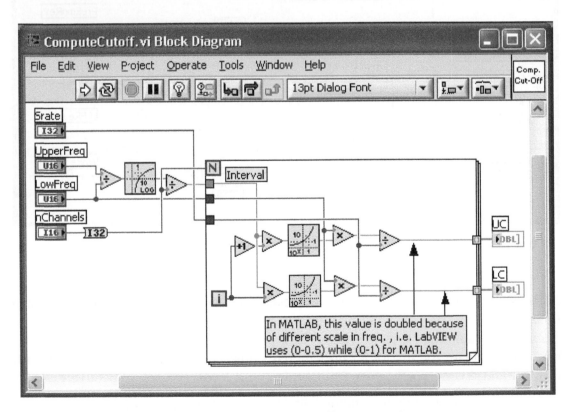

Figure L12-3: Compute Cut-off Frequencies VI.

user-specified input parameters. The inputs `UpperFreq` and `LowFreq` to this VI correspond to the input speech signal bandwidth as specified by f_{low} and f_{high} and the `nChannels` denotes the number of channels n. First, this VI computes the channel bandwidth or interval C_b by using the `Logarithm Base 10` VI (**Functions » Mathematics » Elementary & Special Functions » Exponential Functions » Logarithm Base 10**). The computed channel bandwidth C_b is then passed to a `For Loop` (**Functions » Programming » Structures » For Loop**) along with the sampling frequency `Srate` F_s for computing the cut-off frequencies of the ith channel according to Equations (12.2) and (12.3). To evaluate the cut-off frequencies of the ith channel, one needs to use basic arithmetic VIs together with the `Power Of 10` VI (**Functions » Mathematics » Elementary & Special Functions » Exponential Functions » Power Of 10**). The arrays `UC` and `LC` indicate the upper and lower cut-off frequencies for each of the n channels.

After the cut-off frequencies are computed, they are passed to the `BPFDesign` VI to compute the bandpass filter coefficients of each channel that are used both in the decomposition and synthesis stages. Figure L12-4 shows the BD of the `BPFDesign` VI.

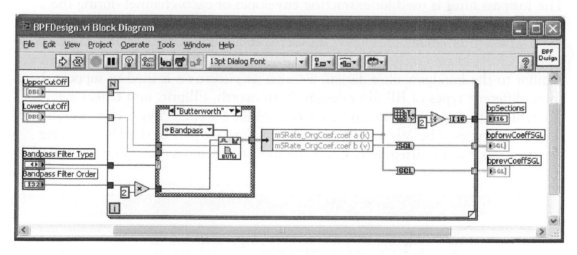

Figure L12-4: Bandpass Filter Design VI.

As indicated in Figure L12-4, a `For Loop` structure is used to compute the bandpass filter coefficients of each channel. The `UpperCutOff` and `LowerCutOff` frequencies obtained from the `ComputeCutoff` VI along with the `Bandpass Filter Type` and `Bandpass Filter Order` are passed to the `For Loop` structure. A `Case Structure` is used to support these three different types of IIR filter design: Butterworth, Elliptic, and Chebyshev. These filters are part of Advanced Filter Design (**Functions » Addons » Digital Filter Design » Filter Design »**

Advanced Filter Design) in the Digital Filter Design toolkit [2]. To achieve this, one needs to use an Enum (**Controls » Modern » Ring & Enum » Enum**) control for Bandpass Filter Type corresponding to these three different filter types. The VIs for IIR filter design that are part of the Digital Filter Design toolkit are utilized here. Once the filter coefficients are obtained in the IIR filter cluster, the forward and reverse coefficients bpforwCoeffSGL and bprevCoeffSGL of each channel are extracted using the Unbundle By Name VI (**Functions » Programming » Cluster & Variant » Unbundle By Name**). Since the filter design using the Digital Filter Design toolkit generates the IIR filter cluster in terms of Second Order Sections (SOS), the BPFDesign VI also computes the number of SOS, bpSections, for any given channel. It should be noted that the UpperCutOff and LowerCutOff frequencies are normalized cut-off frequencies, and the Bandpass Filter Order is thus doubled before passing it to the IIR filter design VIs as per the specifications of the filter design VIs.

L12.1.2 Lowpass Filter Design

The lowpass filter is used for extracting envelopes of each channel during the decomposition stage. Figure L12-5 shows the BD of the LPFDesign VI to compute the lowpass filter coefficients based on user-specified input parameters: Filter Type, LP Filter Order, Cutoff Frequency, and Sampling Frequency. Similar to the bandpass filter design, a Case structure is used to support these three different types of IIR filter design: Butterworth, Elliptic, and Chebyshev. To achieve this, one can use an Enum control for Filter Type corresponding to these filter types. The VIs for IIR filter design that are part of the Digital Filter Design toolkit are utilized here. Once the filter coefficients are obtained in the IIR

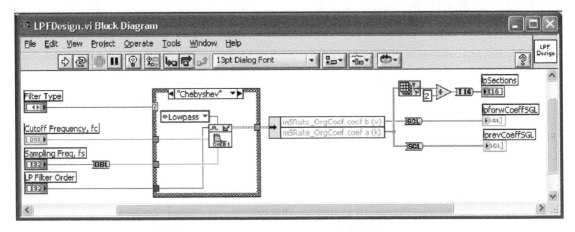

Figure L12-5: Lowpass Filter Design VI.

filter cluster, the forward and reverse coefficients `lpforwCoeffSGL` and `lprevCoeffSGL` are extracted using the `Unbundle By Name` cluster. Since the filter design using the Digital Filter Design toolkit generates the IIR filter cluster in terms of Second Order Sections (SOS), the `LPFDesign` VI also computes the number of SOS, `lpSections`, for the designed lowpass filter.

The outputs of the `LPFDesign` VI and `BPFDesign` VI form part of the outputs of the `DesignFilterCoeff` VI. These outputs, along with the other outputs, namely the pre-emphasis filter coefficients and white noise sequence, are passed to a C DLL, which implements the real-time module. Further details on building the C DLL and its various functions are provided in the following section.

L12.2 Real-Time Implementation

The real-time implementation of the CIS strategy is carried out in a hybrid mode using the `Call Library Function Node` (**Functions » Connectivity » Libraries & Executables » Call Library Function Node**) feature of LabVIEW. This function allows one to invoke C codes via Dynamic Link Library (DLL) and helps to reduce the processing time [3]. Before invoking the functions within a C code, one needs to create a Dynamic Link Library. To create a DLL, use the Visual Studio IDE application. Create a **New » Project** under **File** menu and select **Win32 Project** under the **Visual C++ Project type**. Enter the Project name as **CIS**, choose the option **Create directory for solution**, and enter the Solution name to be the same as Project name, i.e., **CIS**. Now, set the **Application Settings** as **Dynamic Link Library** and choose an **Empty Project**. After creating the Win32 Dynamic Link Library Project solution with the previously mentioned steps, add the source file *CIS.cpp* and the header file *CIS.h* to the project. Before building the project, make sure that the path of the header file is added to **Additional Include Directories** under **Project » Properties » Configuration Properties » C/C++ » General**. Remember that the header file should be included for both the **Release** and **Debug** configurations of the project. Next, choose the **Build » Batch Build** menu and check to build both the **Release** and **Debug** configurations. Once the build is complete, it should create DLLs for both the **Release** and **Debug** configurations in the respective directories under the Project. The *CIS.dll* corresponding to the **Release** configuration is used to configure the `Call Library Function Node` by selecting the appropriate exported function.

Both the source and header files are included in the accompanying CD under the folder `LabVIEW Labs\ Lab12`. In order to avoid the overhead associated with calling the C DLL each time a new speech frame is acquired by using the LabVIEW sound acquisition VIs, the supplied source file *CIS.cpp* utilizes Windows APIs to acquire speech frames from a microphone. It then processes acquired frames based on the noise-band vocoder strategy to obtain synthesized frames and play them back continuously to a speaker using Windows APIs. This is achieved by utilizing the CALLBACK function supported by Windows APIs. Two buffers are used for acquisition and playback so that while one buffer is used for acquiring current samples, the other buffer is used for processing samples.

The source file *CIS.cpp* has a number of functions that perform different tasks. These tasks are summarized in Table L12-1. It should be noted that among all these functions, only the two functions StartCISProcess and StopCISProcess are exported to the C DLL and get invoked using the Call Library Function Node feature of LabVIEW. All the other functions are local to the C code and are invoked internally. Also, it is important to note that the C code for the bandpass and lowpass filtering is generated based on the pseudo code obtained from the Fixed Point tools of the Digital Filter Design toolkit, as indicated in Figure L12-6.

Table L12-1: Building Functions

S.No.	Function Name	Tasks Performed
1.	StartCISProcess	Initializes all filter coefficients and other parameters to global variables and then sets up the CIS real-time loop using CALLBACK function supported by Windows APIs.
2.	StopCISProcess	Stops the real-time loop using the CALLBACK function and then clears all memory allocated to global variables.
3.	InitializeSound	Initializes sound input and output device, that is microphones and speakers. It sets up buffers for speech frames acquisition and playback of synthesized speech. Also, it enables the CALLBACK function to invoke the real-time loop.
4.	CreateMixBuffer	Creates output buffers to write synthesized frames that need to be played back to speakers.
5.	CreateInMixBuffer	Creates input buffers to acquire speech frames from a microphone for further processing according to the CIS strategy and obtains synthesized frames.
6.	WaveInCallbackFunction	CALLBACK function invokes the function ProcessAudioBuffer whenever a new speech frame is acquired and ready to be processed.
7.	ProcessAudioBuffer	Processes an acquired input speech frame according to the CIS strategy, which includes both the decomposition and synthesis stages and finally obtains synthesized speech for playback onto speakers.

Table L12-1: Building Functions—Cont'd

S.No.	Function Name	Tasks Performed
8.	DestroySound	Closes input and output devices that were opened using Windows APIs. It also clears all memory allocated for global variables.
9.	PreEmphasis	Performs the pre-emphasis filtering to obtain pre-emphasized output.
10.	FullWaveRect	Obtains the full-wave rectified output of an input signal.
11.	BandpassFilteringFP	Initializes the previous states of each second order section of the designed bandpass filter for a given channel and then invokes the function BPFilter_FilteringFP to filter each input sample of current speech frame.
12.	BPFilter_FilteringFP	Obtains a corresponding bandpass filter output for a given input sample by passing through a cascaded stages of second order sections.
13.	LowpassFilteringFP	Initializes the previous states of each second order section of the designed lowpass filter for a given channel and then invokes the function LPFilter_FilteringFP to filter each input sample of current speech frame.
14.	LPFilter_FilteringFP	Obtains the corresponding lowpass filter output for a given input sample by passing through a cascaded stages of second order sections.

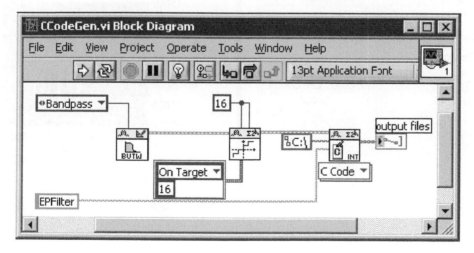

Figure L12-6: Generate C code using Fixed-Point tools of DFD toolkit.

As indicated in Figure L12-6, the designed floating-point filter coefficients are first quantized to generate a fixed-point filter using DFD FXP Quantize Coef (**Functions » Addons » Digital Filter Design » Fixed-Point Tools » DFD FXP Quantize Coef**), and then DFD FXP C Code Generator (**Functions » Addons » Digital Filter Design » Fixed-Point Tools » DFD FXP Code Generator**) is used to generate the C code. When the VI is executed, it creates the pseudo code for fixed-point filtering in C and stores the files in the specified path. One can locate the source file created and modify the fixed-point filtering code in C to floating-point filtering accordingly— that is, by changing the input and output data types and removing the scale factors. It should be noted that the C code thus generated implements the filtering operation using cascaded stages of second order sections.

Once the DLL is created, the functions are exported via the Call Library Function Node feature in LabVIEW, as shown in Figure L12-1. As indicated in this figure, one of the Call Library Function Nodes is configured to invoke the exported function StartCISProcess, whereas the other is configured to

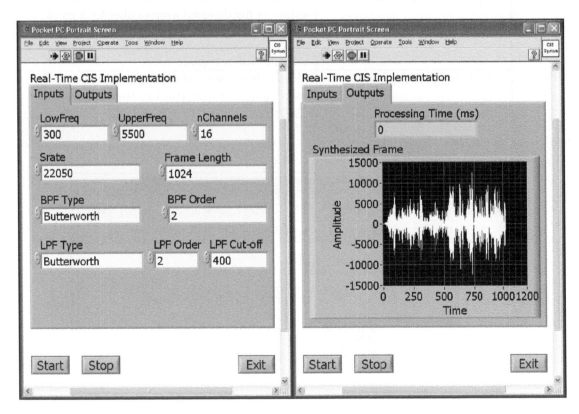

Figure L12-7: Interactive FP of cochlear implant simulator.

invoke the exported function `StopCISProcess`. In order to configure any `Call Library Function Node`, right-click on the `Call Library Function Node` and choose the **Configure** option. Then choose the path of the C DLL, select the exported function, and add the appropriate input and output parameters as per the definition of the exported function.

As can be seen from Figure L12-1, all the relevant inputs are passed to the C DLL via the `Call Library Function Node`. When the user presses the `Stop` Boolean control, the synthesized frame last obtained and the `Processing time (ms)` are displayed; see the simulator FP displayed in Figure L12-7. It should be realized that the pointers to the output variables are passed to the C DLL, and then the output is updated within the C DLL accordingly. The FP of the simulator allows the user to interactively change various input parameters and then observe the corresponding change in outputs. When the user presses the `Start` Boolean control, the synthesized frame is played back to the speakers in response to the speech through a microphone and halts the program either when the `Stop` Boolean control or the `Exit` control is pressed.

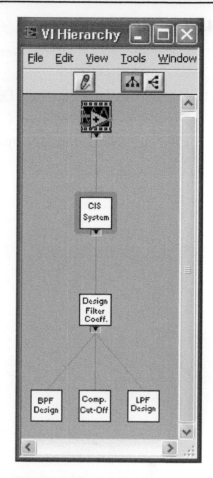

Figure L12-8: Hierarchy of cochlear implant simulator VI.

In summary, a cochlear implant simulator system is built in hybrid mode. A simplified version of the system hierarchy, displayed by choosing **View » VI Hierarchy**, is shown in Figure L12-8. When C DLLs are used, the real-time implementation of the CIS strategy, which includes the decomposition and synthesis stages, is achieved. All the subVIs discussed in this lab can be saved in a LabVIEW Library (LLB) file, e.g., *Lab 12.llb*. A new LLB file can be created by choosing **New VI Library** and naming it from the **Name the VI** window, which is brought up during the save operation.

L12.3 Bibliography

[1] Microsoft Developer Network, *Visual* C++, MSDN Library Visual Studio, 2007.

[2] National Instruments, *Digital Filter Design Toolkit* (search keywords), http://www.ni.com.

[3] National Instruments, *Integrating C DLLs* (search keywords), http://www.ni.com.

Index

A

acoustic signals, 303
adaptive filtering systems
 DSP integration of, 248
 system identification of, 248–251
adaptive FIR filter, usages of, 157
A/D converter, 62
 bitstream of, 84
 characteristic of, 63
 data quantization effect, 127
 quantization effect, analog signal, 80
  output signal, 82–84
   quantization error, 81–83
 quantization noise, 63
additive quantization noise, 63
Add Zeros VI, 88
aliasing, of discrete time signal, 60
analog signal
 processing of, 1
 sampling of, 57
analog-to-digital signal conversion
 quantization, 63–64
 sampling, 57
 signal reconstruction, sinewave, 65–66
 Sine Wave VIs for
  aliased signal, 72
  inputs to, 71
application programming interfaces
  (APIs), 229
array
 creation using auto-indexing, 33–34
 of data elements, 16
Array Subset function, 116
auto-indexing, 33
 input array, For Loop, 86

B

Bandpass filters, 304, 306
Bandpass IIR filter. *See* IIR filter
bandwidth, 279
baseband signal, 279
Basic Function Generator VI, 46
binary bitstream, 86
binary point, 124

Block Diagram (BD)
 of aliasing, 72–73
 building
  Express VI and function, 11–12
  structure, graphical enclosure, 13–14
  terminal icon, 12
  wires, data transfer, 12–13
 control and indicator icons, 30
 of Express VI FFTs, 76
 of FFT and STFT, 187
 of noise cancellation system, 168
 of signal generation and amplification
   system, 38
 of signal reconstruction system, 88
 Subtract function on, 163
 of system identification, 164
 wiring of, 24
boolean control, 309
breakpoints, 16, 205, 216–218, 220–221
Building C DLL, 51–52
Build Options of CCS, 236
Butterworth Filter VI, 248
Butterworth IIR filter, 121

C

CALLBACK function, 316
Calling C DLL, 52–54
Call Library Function
 building C DLL, 51–52
 calling C DLL, 52–54
 configuration of, 54
Call Library Function Node VI, 119
cascade coefficient cluster, 111–112
cascaded filter, stages of, 96
Cascaded integrator comb (CIC), 98
cascade-form coefficients, 112
case selector, 14
Case Structure, 14
CCS Automation, 231
CCS Build VI, 226
CCS Communication VIs, 235
CCS Download Code VI, 226
CCS Open Project VI, 226
CCS Reset VI, 226

CCS RTDX Read, 234
CCS RTDX Read Array SGL VI, 235
CCS RTDX Write Array SGL, 234
CCS RTDX Write VI, 234
CCS Run VI, 226
C DLL, 309
central processing unit (CPU), 197
C function, process for adding, 218
channel noise, 297
Chebyshev IIR filter, 121
Chebyshev method, 94
Chirp Pattern VI, 186
CIC. *See* Cascaded integrator comb
CIS.cpp, 315–316
C Langauge, filtering code written in, 233
cluster control, 269
clusters, 16
cochlear implant simulator vi, hierarchy
  of, 319
cochlear implant system
 components of, 303
 filter design
  bandpass filter design, 312–314
  DesignFilterCoeff VI, 310–312
  lowpass filter design, 314–315
 interactive FP of, 318
 noise-band vocoder simulation of, 304
 system-level VI of, 310
Code Composer Studio (CCS), 204–205,
   207, 223
 automation process for, 226,
   231–233
 RTDX communication, 227
code execution, time profiling using
   breakpoints for, 221
coefficient quantization, 128
coefficient quantization, 128
compiler options, 212
Compiling, 205
Composite Signal VI, 186
conditional terminal, 14
configuration dialog box, 77
connector pane, of LMS VI, 163
Continuous Interleaved Sampling (CIS)
  real-time implementation of, 305

This book combines textual and graphical programming to form a hybrid programming approach, enabling a more effective means of building and analyzing DSP systems. The hybrid programming approach allows the use of previously developed textual programming solutions to be integrated into LabVIEW's highly interactive and visual environment, providing an easier and quicker method for building DSP systems. This book will be an ideal introduction for engineers and students seeking to develop interactive DSP systems quickly.

Features

- The only DSP laboratory book that combines textual and graphical programming

- 12 lab experiments that incorporate C/MATLAB code blocks into the LabVIEW graphical programming environment via the MathScripting feature

- Lab experiments covering basic DSP implementation topics including sampling, digital filtering, fixed-point data representation, and frequency domain processing

- DSP system applications using the hybrid programming approach, such as a software-defined radio system, and a cochlear implant simulator system

- CD providing all the lab codes

About the Author

Nasser Kehtarnavaz is Professor of Electrical Engineering at University of Texas at Dallas. He has written numerous papers and six other books pertaining to signal and image processing, and regularly teaches digital signal processing laboratory courses, for which this book is intended. Among his many professional activities, he is Coeditor-in-Chief of *Journal of Real-Time Image Processing* and Chair of the Dallas Chapter of the IEEE Signal Processing Society. Dr. Kehtarnavaz is a Fellow of SPIE, a Senior Member of IEEE, and a professional engineer.

Printed and bound by CPI Group (UK) Ltd, Croydon, CR0 4YY

03/10/2024

01040310-0010